The Prince Reimagined

Also from EATMS Productions

Books on power, survival, women's autonomy, and the systems shaping modern America.

Nonfiction

Billionaires, Capitalism, and Power

Evil and the Mountain Ungreed
Self Help for American Billionaires
Selfish Steve and the Ivory Tower
Tariffs, Taxes, & Face-Eating Leopards
Ban Billionaires: Fascism Fix

Fascism, Religion, and Cultural Control

Self Help for the Manosphere
Fascism 2025
Fascism & the Perverts & the Greed Virus
Christian Fascism Marriage Book
Tyranny, Table Manners, & Tiramisu

Guides for Women's Autonomy and Protection

How to Survive in Post-America as a Woman
Project 2025 American Drag
4B – Burn, Ban, Boycott, Build
4B OG – So No Go GYN
I'm Glad He's Dead

Analysis of Authoritarian Project 2025

Project 2025: The Blueprint
Project 2025: The List
Project 2025, Christian Dumb Dumbs, & The Republican Agenda
Fascism, Project 2025, & The Pinkprint

Modern Rewrites for Women

Stoic Principles Reimagined
Siddhartha Reimagined
The Prince Reimagined for Women
The Art of War Reimagined for Women
The Jungle Reimagined
The Constitution Reimagined for Women

Machine Learning Series

AI, Bitcoin, Nostr for Women
AI, Safety, & Security for Women
AI, Anxiety, & Health for Women
AI, Kids, & Family Safety for Women
AI, Creativity, & Personal Expression for Women
AI, Independent Work, & Parallel Power for Women

Social Systems Series

Emotional Labor for Women
Household Power for Women
Workplace Power for Women
Medical Bias for Women
Aging Systems for Women
Recovery Systems for Women

Fiction

Dystopian Stories of Resistance and Collapse

Propaganda Paige & the Missing Prosperity
Propaganda Paige & the TIDE Manifesto
Propaganda Paige & the Shadow Cartographers
Propaganda Paige & the Prosperity Alliance
Propaganda Paige & the Shattered Truth
Propaganda Paige & the Rising TIDE
Propaganda Paige & the Last Bastion
Propaganda Paige & the Dawn of Prosperity
Project 2025: Dorian — The Last Men
Project 2025: Boy — A Last Men Novel

The Prince Reimagined for Women 2025

Sympathy & Support 2

Inspired by Niccolò Machiavelli's
Original Version of The Prince

by
Esme Mees
& Claudia Mayfair

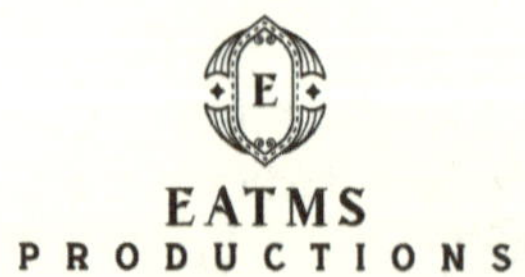

This title is part of an ongoing body of work. All EATMS Productions titles, across all series, authors, and formats, are components of a single connected project.

This book is a work of opinion and creative interpretation. While some names and events may be referenced or alluded to, any claims made are based on publicly available information and are intended as satire, parody, or commentary on societal and political issues. The content should not be interpreted as factual assertions about any individual or entity. The author does not intend to defraud, defame, or mislead, and encourages readers to form their own conclusions. Any resemblance to real persons, living or dead, is purely coincidental unless explicitly noted otherwise.

ISBN 978-1-966014-12-6

Cover, interior design, interior prints by: Esme Mees

eatms@pm.me
www.eatms.me

Check out EATMS Underground:
https://tinyurl.com/eatmsNOSTR

Printed in the United States of America

The vulgar are always taken by what a thing seems to be.

—Niccolò Machiavelli, *The Prince*

Table of Contents

Introduction

The Reality of Power

Power is never evenly distributed. It is deliberately concentrated, protected by those who benefit from its imbalance, and wielded with intent. It is seized, maintained, defended, and, more often than not, ruthlessly protected by those who have it. Niccolò Machiavelli, writing in Renaissance Italy, sought to instruct rulers on the mechanics of control, the necessities of deception, and the cold pragmatism required for survival in a world that is neither fair nor kind. But his audience was never meant to include women. His lessons on power, fear, and manipulation were crafted for men who were assumed to be the natural wielders of influence. Women, if they appeared at all in these calculations, were either obstacles to be managed or assets to be married off in service of dynastic ambitions. Five hundred years later, the world has changed, or so we like to tell ourselves. The truth is more complicated. Power remains a battlefield, and women, despite all supposed progress, are still expected to fight for what men take as their birthright.

To understand Machiavelli's relevance in 2025, one must first recognize that the fundamental principles of power have remained unchanged. Power is still dictated by control over resources, the ability to command influence, and the necessity of strategic maneuvering. What has changed is the illusion of progress. While women have gained access to power in ways that would have been unthinkable in Machiavelli's time, the structures that dictate power have merely adapted to mask their barriers rather than remove them. Women are still penalized for ambition. They are still forced to prove themselves twice over, to navigate spaces where their presence is tolerated but not truly welcomed. In politics, business, and even personal relationships, the struggle for power is fraught with additional expectations. Machiavelli wrote for rulers who were assumed to be legitimate in their authority. Women, on the other hand, are rarely given that assumption. They must justify their leadership at every turn.

The contemporary battlefield is not one of sword and conquest but of systemic constraints, media narratives, and the silent but powerful machinery of bias. The perception of female power remains precarious, women must project authority without being seen as overbearing, must negotiate without appearing weak, and must command respect without triggering resentment. A woman in power does not simply lead; she is constantly performing a balancing act. Machiavelli understood that power is often maintained through perception rather than reality. This is particularly true for women, who must carefully curate their public image while their male counterparts are free to govern as they see fit. A man can be ruthless and be admired for his decisiveness; a woman who does the same is branded cold and unlikable. The battlefield remains uneven, and the weapons of warfare have evolved.

The first rule of power for men is that it is assumed. The first rule of power for women is that it must be explained. This distinction shapes every interaction, every leadership decision, and every attempt at influence. A male politician can present himself as a leader with little concern for likability; a female politician must carefully craft an image of strength without alienating voters who still expect women to be nurturing. A male CEO can be direct and aggressive; a female CEO who does the same is often criticized for lacking emotional intelligence.

This expectation forces women to operate on two levels: the practical and the performative. They must execute the responsibilities of leadership while simultaneously managing perceptions, ensuring that their authority is palatable to those who instinctively resist it. This is not a burden placed on men, who are permitted to be flawed, to make mistakes, to wield power without constant scrutiny.

The disparity in how men and women experience power is evident in every field:

- *Politics:* Male politicians are judged on policy and effectiveness; female politicians are judged on likability, wardrobe, and the tone of their voice. The threshold for competence is higher, the tolerance for error lower.
- *Corporate Leadership:* Men in executive roles are expected to lead with confidence and decisiveness; women in the same roles must tread carefully between being assertive and being perceived as aggressive.
- *Media Representation:* Male public figures are afforded complexity; female public figures are reduced to tropes, too emotional, too harsh, too ambitious. Their authority is questioned in ways that male authority never is.

The justification of female authority is a never-ending performance. Women must convince others, and often themselves, that they belong in positions of power. They must anticipate resistance and counteract it before it takes hold. They must walk the tightrope of public perception, knowing that a single misstep will be judged more harshly than if a man had made the same mistake. This performance is exhausting, but it is necessary. The game is unfair, but the only way to win is to play it better than those who have written the rules. And in playing it better, women must also work to change the rules entirely.

If power is a battlefield, then the strategy must evolve. Women cannot simply integrate into the existing structures and expect change; they must dismantle and rebuild them. This means shifting the conversation from justifying female leadership to normalizing it. It means rejecting the expectations that have been placed upon women for centuries and refusing to be constrained by them. Machiavelli's lessons remain relevant because they expose the nature of power as it truly is, unsentimental, strategic, and often brutal. For women, the challenge is even greater because the opposition is not only external but deeply embedded within societal norms. However, history has shown that women who recognize these constraints, who master the strategies necessary for survival, and who refuse to wait for permission, can and will reshape the world.

Power has never been a neutral terrain. It is designed, structured, and hoarded by those who already possess it, and when women seek entry, the gates do not open willingly. The barriers women face, political, financial, cultural, are not incidental but deliberately maintained. They are not accidents of history but artifacts of deliberate exclusion, reinforced over centuries to keep power concentrated in the hands of men. Women who wish to wield authority must understand the nature of these barriers, not as abstract inconveniences but as engineered systems meant to exhaust, deter, and ultimately deny them access.

Niccolò Machiavelli's *The Prince* was written in a time when power was a physical, territorial conquest. But in the modern world, power is embedded within institutions, laws, and economic hierarchies. To dismantle these structures, women must become fluent in their mechanisms, how they are upheld, who benefits from them, and where the points of fracture lie. The first lesson: waiting for systemic change is a fool's errand. No one cedes power voluntarily. If women want a different world, they must take it.

Women are told that democracy is an equal playing field, that participation is as simple as running for office, organizing, voting. But the moment a woman steps forward, she encounters a reality far removed from the fantasy of democratic inclusion. In politics, women must first prove they are *qualified*, an absurd expectation when unqualified men ascend to the highest offices without question. They must be likable but not weak, authoritative but not threatening. If they show ambition, they are called power-hungry; if they hesitate, they are seen as incompetent.

The political system is designed not just to exclude women but to break them in the process of entry. The relentless scrutiny, the double standards, the media's obsession with their appearance rather than their policies, all of it serves as a deterrent. And yet, women still fight their way in. The question is not whether they should participate but how they can subvert a system built against them. A woman in politics must accept that she will be vilified, underestimated, and attacked. The solution is not to seek approval from those who will never grant it but to redefine the game. Women must enter political spaces not as guests seeking permission but as insurgents determined to seize control. The system will not change from within unless the women who enter it intend to burn its old structures to the ground.

History offers ample evidence that political power is not given; it is wrested through force, strategy, and resilience. Consider Queen Elizabeth I, who ruled England against all odds. She understood that to survive, she had to outmaneuver her male advisors, avoid marriage (which would have placed her under a man's control), and cultivate an image of divine authority. Or take Cleopatra, whose political acumen and alliances with Rome allowed her to rule Egypt even as foreign and domestic enemies sought her downfall. These women did not ask to be included; they seized power with calculated precision.

The lesson for modern women? Do not expect the system to play fair. Understand its rules better than those who wrote them. Use its mechanisms against it. Women in politics today must master the art of leverage, aligning with power where necessary but never losing sight of their own autonomy. Power is not a favor; it is a prize for those bold enough to claim it.

If politics is the stage where power is performed, finance is the shadowed engine that controls the script. Women are encouraged to work hard, to seek financial independence, to break through glass ceilings. But what is never acknowledged is that the very structure of economic power is designed to keep them at a disadvantage. The wage gap, the investment gap, the near-total male dominance of venture capital and corporate leadership, these are not accidental disparities. They are systemic levers of control.

Consider this: women-owned businesses receive less than 2% of venture capital funding. Women are penalized for negotiating salaries, while men are rewarded for the same behavior. Financial literacy is not taught as a survival skill for young women because keeping them uninformed ensures their continued subjugation. Even when women earn wealth, they are socially conditioned to feel uncomfortable wielding it, as if financial power is an unnatural extension of their ambition.

The first step in dismantling financial barriers is rejecting the myth of meritocracy. Hard work does not guarantee success in a rigged system. The next step? Women must build their own financial power networks, independent of the institutions that have historically shut them out. This means investing in each other, creating alternative funding sources, and refusing to participate in economic structures that exploit them. Money is not just currency; it is power. Women must stop seeing it as a reward and start treating it as a weapon.

Throughout history, women have found ways to subvert financial barriers, even when the law explicitly denied them control over wealth. In the 19th century, American heiresses like Hetty Green wielded financial influence in a world where women were legally barred from managing their own money. Medieval European noblewomen often acted as the de facto rulers of their estates when their husbands were absent, proving that economic power was never a question of ability but of restriction. The modern lesson is clear: financial power is not about earning, it is about controlling resources. Women must claim ownership over economic structures rather than simply participating in them. This means not only earning wealth but directing where it flows, who benefits from it, and how it is used to create systemic change.

Perhaps the most insidious barrier is cultural, the deeply embedded belief that women in power are unnatural, dangerous, or destined for downfall. Every society crafts myths that justify its power structures, and the most enduring myth is that female authority is a threat. From religious dogma that frames women as subservient to the endless portrayals of ambitious women as villains in literature and media, the message is clear: women who seek power must be punished. This cultural conditioning is why women who rise are met with immediate backlash. It is why female leaders are scrutinized for their emotions, their appearance, their personal lives, while men are evaluated solely on their actions. It is why women in leadership are held to impossible standards, expected to be both nurturing and ruthless, competent yet deferential, powerful but never intimidating.

The strategy to dismantle this barrier? Own the narrative. Women must redefine leadership on their terms, refusing to conform to expectations designed to keep them constrained. They must embrace the reality that they will be called difficult, aggressive, or even monstrous, and accept these labels not as insults but as indicators that they are truly disrupting the status quo. History's most powerful women were almost always rebranded as villains. Empress Wu Zetian of China was a formidable ruler, yet she was painted as a ruthless usurper. Catherine the Great expanded Russia's empire and modernized its institutions, yet she was slandered with outlandish rumors meant to discredit her authority. These women understood that power does not accommodate; it demands. Modern women must take the same approach. The goal is not to be liked, it is to be effective. Power is never about fitting into a mold that was never designed for them. It is about breaking it apart and building something entirely new.

The barriers women face, political, financial, cultural, are not insurmountable, but they are deeply entrenched. Dismantling them requires more than participation; it requires strategy, defiance, and an unrelenting willingness to claim power rather than wait for it. Women who understand the game must not merely play it, they must change it. Power will never be granted freely. It must be taken, wielded, and defended. And the women who dare to do so must accept that they will be feared, reviled, and, ultimately, unstoppable.

The Machinery of Power

Power in the 21st century is a machine, finely tuned to preserve itself. It is not wielded by democratically elected representatives or the carefully marketed "leaders" that institutions parade in front of the public to sell the illusion of inclusivity. It is hoarded, protected, and manipulated by the billionaire class, men who have realized that control over wealth is control over the future itself. And when women seek to enter this power structure, whether politically, financially, or culturally, they are met with swift and calculated opposition. The conservative backlash against women's rights, the suppression of female leadership, and the mirage of progress are all symptoms of the same disease: a power system that is allergic to disruption.

While history offers countless examples of women who managed to wield authority despite opposition, their stories are often rewritten, sanitized, or buried altogether. The lesson of their survival is not in their compliance but in their audacity. Today's fight for power is no different, and women must recognize that they are up against something far more sophisticated than individual misogyny or outdated prejudices. They are up against a system that thrives on their exclusion.

The billionaire class has created a world where money does not just buy influence, it manufactures reality. These men are not content to merely dominate industries; they dictate policies, control media narratives, and decide which political movements live or die. They have turned democracy into a farce, funneling dark money into elections, stacking courts with judges who will protect their interests, and ensuring that the illusion of choice remains intact while the outcomes are already predetermined. Women's rights are often the first sacrifice on this altar of greed, because empowering women disrupts their ability to dictate the rules of power. Conservative billionaires pour unfathomable sums into think tanks, media outlets, and political campaigns designed to push a regressive agenda under the guise of "traditional values."

They do not believe in these values, they believe in power, and they understand that controlling women's autonomy is a fundamental part of maintaining it. Every rollback of reproductive rights, every attack on gender equity in the workplace, every insidious attempt to repackage male dominance as "family values" is a strategic maneuver, not an accident. They know that an independent, financially secure woman is a threat to the structures they have spent generations constructing, because she cannot be controlled as easily as the mythologized, dependent, obedient version of womanhood they want to enforce.

The illusion of progress is one of the most effective weapons in this battle. Women are constantly told how far they have come, as if the mere presence of female CEOs, politicians, or celebrities is evidence that power has been equitably distributed. But for every highly publicized female executive, there are millions of women struggling against a system that refuses to pay them equally, that denies them access to leadership, that punishes them for the same ambition that rewards men. For every woman who wins an election, there are countless more who are pushed out, sabotaged, or discouraged from running in the first place. Progress is not about symbols or representation, it is about structural change, and structural change is precisely what the ruling class fears the most. They will celebrate the optics of women's success while ensuring that the mechanisms of power remain untouched.

This is why conservative backlashes are cyclical; as soon as women make tangible gains, there is an immediate and forceful effort to roll them back. The victories are allowed only to a point, and once that point threatens the hierarchy itself, the reaction is swift. We have seen this in every major struggle for women's rights, from suffrage to workplace equity to reproductive freedom. The moment a movement gains momentum, the machinery of opposition mobilizes: billionaires fund reactionary campaigns, media outlets begin pushing narratives about how feminism has "gone too far," and laws are rewritten to remind women that their autonomy is conditional.

But history is full of women who refused to accept these conditions, who understood that power is not something granted, it is something seized. Elizabeth I of England was never meant to rule; she was a last resort, a woman tolerated only because there were no male alternatives. Yet she transformed England into a global power, using political maneuvering, intelligence, and sheer force of will to outlast her enemies. She refused to marry, knowing that doing so would make her a puppet, and instead built an image of divine authority that made even the most skeptical men bow before her. Catherine the Great, a foreigner and a woman, took control of Russia not through inheritance but through a coup, unseating her husband and reshaping the empire with reforms that outlived her. These women did not play by the rules set for them; they rewrote the rules entirely. They understood that waiting for permission to wield power is a fool's game, one designed to keep women in a state of perpetual subordination.

Even in more recent history, we see women who broke through systemic opposition with strategies that mirrored Machiavelli's teachings. Angela Merkel was underestimated at every stage of her career, dismissed as uncharismatic and unlikely to last. Yet she became one of the longest-serving and most effective leaders in modern European history, outmaneuvering both domestic and international opponents with a pragmatism that left her male counterparts scrambling to keep up. Indira Gandhi, despite the dynastic expectations placed upon her, ruled India with an iron grip, making decisions that were often ruthless but undeniably effective in consolidating her power. These women understood that power is never gentle, that to maintain authority in a world designed to strip them of it, they had to be harder, smarter, and more strategic than the men around them.

The lesson from both past and present is clear: women who succeed in power do so not by appealing to fairness or hoping for systemic change, but by understanding the brutal mechanics of control. The modern billionaire class and the conservative backlash it funds are not concerned with moral debates or ideological purity, they are engaged in a war for dominance, and they play to win. Women must adopt the same mindset. This does not mean embracing cruelty or abandoning principles, but it does mean recognizing that power is not about who deserves it, it is about who takes it. The opposition to women's power is not rooted in logic or fairness; it is rooted in fear. Fear that when women gain control, they will upend the carefully balanced hierarchy that keeps wealth, influence, and decision-making in the hands of a select few. And that fear is justified.

The fight ahead is not just about representation or incremental change; it is about dismantling the very structures that uphold male dominance. Women must stop waiting for the world to become more welcoming. They must take what they need, build networks that are independent of male-controlled institutions, and refuse to be placated by the hollow gestures of progress that the ruling class offers in place of real power. There is no reason to play fair in a system designed to keep them losing. Women must embrace strategy over sentiment, pragmatism over politeness, and the undeniable truth that power is never handed over, it is seized. The billionaires, the reactionaries, and the systems that sustain them all understand this. It is time women did too.

Esme Mees & Claudia Mayfair, Winter 2025

~1
On Power

Why Women Must Take It, Not Wait for Permission

Power is never freely given. It is not handed over in an act of goodwill, nor is it bestowed upon those who wait patiently, hoping that merit alone will carry them forward. The myth of waiting, the idea that power is something to be earned through compliance, through good behavior, through perseverance without confrontation, is one of the most dangerous fictions women have been taught. It is a fairy tale, one designed to keep them docile, manageable, and permanently excluded from the real mechanisms of control. The world does not reward passivity; it exploits it. Women who wait for power to be granted to them will die waiting, watching as less competent men ascend, as systems reinforce themselves, as doors open only for those who have never been forced to knock. Power is seized, taken with intention, defended with strategy, and maintained with ruthless pragmatism. The moment a woman understands that, she stops asking for permission.

The reality is that every structure of authority, political, economic, cultural, has been engineered to exclude women, or at the very least, to manage them. And yet, despite the promises of progress, despite the language of equality now woven into the rhetoric of corporations, media, and politics, the essential nature of these structures remains unchanged. Women are told they are empowered while they are actively denied real access to power. This is the modern sleight of hand, the illusion that keeps women believing in a system that was never designed to serve them. If women wish to hold power, they must understand the battlefield they are stepping onto. They must abandon the fantasy of fairness, reject the expectation of reward for patience, and recognize that their only path forward is the one they carve for themselves.

The corporate world is perhaps the most insidious manifestation of this illusion, where women are paraded as symbols of progress while being systematically obstructed from true leadership. They are celebrated for breaking "glass ceilings" that remain firmly in place, their individual successes turned into marketing campaigns to obscure the larger reality. Women are still paid less, promoted less, and given fewer opportunities to advance, and yet every high-profile female executive is used as proof that the problem has been solved. The truth is that corporations have no interest in dismantling male-dominated hierarchies. They have every interest in maintaining them, while selling the appearance of change. The language of "empowerment" is a branding strategy, not a promise. It is meant to pacify, to keep women striving within a rigged system rather than dismantling it altogether.

Political gatekeeping operates under a similar pretense. Women are encouraged to run for office, to participate, to engage, so long as they understand the unspoken rules of the game. They must be competent, but not threatening; assertive, but not aggressive; likable, but not weak. If they step too far outside the bounds of what is deemed acceptable, they are vilified. The scrutiny is relentless, the double standards suffocating. And yet, even the most obedient, the most calculated, the most strategically palatable women still find themselves excluded from real decision-making. The message is clear: play by the rules all you want, but the game is still designed for you to lose.

It is at this point that history becomes instructive. The women who have successfully wielded power did not do so by waiting for approval or by conforming to expectation. They understood that power is neither inherited nor bestowed, it is taken. They knew that rebellion was often the only path forward, that strategy required an ability to see beyond the immediate moment, that force was sometimes necessary when all other avenues were closed. Elizabeth I did not survive as queen by hoping her councilors would accept her authority. She ruled with a calculated balance of charm and ruthlessness, refusing to be controlled by marriage, manipulating the perception of her reign to ensure that her legitimacy could not be questioned. Catherine the Great did not politely request to rule Russia; she staged a coup against her own husband and took the throne by force. These women, and many others like them, recognized that waiting was not an option. They did what was necessary.

Even outside the realm of monarchy, women have had to seize power in ways that defy expectation. Women-led revolutions have rarely been granted the recognition they deserve, but history is full of them. From the women who orchestrated resistance movements in times of war, to those who defied social conventions to build independent wealth and influence, the pattern is always the same: power came only when it was demanded, not requested. In the French Revolution, women took to the streets, wielding both weapons and words to demand their rights, only to be excluded from the very political structures they helped create. In the labor movements of the early 20th century, women organized strikes, shut down factories, faced imprisonment, and yet, even when their victories reshaped economies, they were often written out of the narratives of change.

The lesson is simple but brutal: power does not accommodate. It does not bend to politeness. It does not reward those who ask nicely. It is taken by those who refuse to accept exclusion. This does not mean abandoning ethics or integrity, but it does mean understanding that the world is not fair, and that fairness should not be expected. Machiavelli, when writing *The Prince*, did not concern himself with what was just, he concerned himself with what was effective. And for women, effectiveness means recognizing that every barrier in front of them is not a test of patience, but a challenge of will.

The modern woman must operate with the same clarity. She must recognize that the structures meant to suppress her are not accidents, but intentional obstructions. She must stop accepting the false narratives of empowerment that corporations and politicians use to keep her satisfied with symbolic progress. She must reject the idea that power is something that will come to her if she simply waits long enough. Instead, she must take it, by force, by strategy, by relentless persistence. She must recognize that opposition is inevitable, that backlash is certain, that those in power will always attempt to maintain their dominance. But she must also understand that fear of opposition is the very thing that keeps power concentrated in the hands of those who least deserve it.

The path forward is not about seeking approval from a system that was never built for women's success. It is about dismantling that system and replacing it with something better. Women must build their own networks of power, bypass the gatekeepers who refuse to let them in, and redefine leadership on their own terms. They must understand that they will be called aggressive, unlikable, too ambitious, too demanding, and they must learn to wear these labels without shame. Because the alternative is submission, and submission is not an option. Women who wait do not win. Women who take, who demand, who insist, those are the women who shape the world.

This is not a call for recklessness, but for strategy. Power is not acquired through hope; it is built through calculated action. Women must study the structures they wish to infiltrate, learn the weaknesses of those who stand in their way, and recognize that every gain requires confrontation. They must understand that power is not about fairness, it is about leverage. And they must be willing to embrace the reality that their victories will not be celebrated by those who wish to see them fail. The world will not congratulate women for taking power. It will fight them every step of the way. But that fight is not a reason to back down, it is the very reason they must push forward.

The time for waiting is over. The time for politeness is past. Women must not ask for power. They must take it.

Part 2

Power is an act of will, not a gift. It is not something given in reward for patience or good behavior, nor is it a prize distributed based on merit. Power exists to be seized, and those who hesitate, who wait for fairness, who believe the rules were ever meant to include them, will find themselves left behind, outmaneuvered, controlled. Women have spent centuries being told to wait their turn, to prove their worth, to climb ladders that lead nowhere, and yet history has been shaped by those who ignored these commands. The illusion of progress, the carefully curated narratives of inclusion, the symbolic victories meant to pacify rather than empower, these are all tactics of those who understand that real power is about control, and control is never handed over voluntarily. Women who wish to lead must first recognize that they will never be welcomed into power; they must break into it.

The corporate world thrives on the myth that women are advancing, that the doors are open, that those who work hard enough, lean in far enough, will eventually be allowed to take their place at the top. But this is a mirage, designed to keep women investing in a system that continues to deny them. Corporate structures do not exist to be fair; they exist to maintain hierarchies, and those hierarchies are built on exclusion. The so-called "glass ceiling" is not a flaw, it is a design feature. When a woman does manage to ascend, she is often isolated, used as proof that the system works, paraded as evidence that there are no longer barriers. Meanwhile, the structures remain intact, the wages remain unequal, the decisions remain controlled by the same hands that have always controlled them.

Women in these spaces must stop playing by rules meant to keep them contained. They must stop believing in the idea that their competence alone will be enough. Competence is not what determines power. Men do not dominate boardrooms and governments because they are the most capable, the most intelligent, the most qualified. They dominate because they have no hesitation in claiming what they believe is theirs. Women must cultivate the same lack of hesitation, the same expectation of authority, the same refusal to be sidelined. This means taking power where it is not offered, stepping into roles that were not designed for them, and building their own systems where necessary. It means recognizing that every barrier is intentional, and that the only way past it is through force, whether that force is financial, political, or social.

Politics operates under the same façade of progress. Women are encouraged to run, to participate, to believe that their voices matter, but only as long as they do not truly threaten the established order. The moment they do, the system reveals its true face. The attacks come swiftly: media scrutiny, character assassination, the framing of ambition as corruption. They are labeled unelectable, too emotional, too cold, too weak, too strong. The contradictions are deliberate. They are not meant to be resolved; they are meant to exhaust. If a woman remains in the race despite all this, she is met with further barriers, gerrymandering, financial obstruction, networks that refuse to support her. And yet, even as women are pushed out, the narrative remains the same: progress is happening. Women are told to celebrate the few victories, to be grateful for incremental change, to keep believing that equality is just beyond the horizon. But it is not. The horizon only moves further away the longer women chase it.

The women who have held power throughout history have understood this. They did not wait for permission, nor did they waste time trying to conform to the expectations placed upon them. They broke rules, they defied norms, they forced their way into positions they were never meant to hold. Take Cleopatra, who ruled Egypt not as a placeholder but as a sovereign force, maneuvering between Rome's power struggles, using every tool available to her, political alliances, military strategy, economic control, to maintain her position. She was not a ruler because she was allowed to be one. She was a ruler because she refused to be anything else. The same can be said for Empress Matilda, who fought for her rightful throne in 12th-century England against a system that refused to acknowledge her legitimacy. She led armies, she outlasted betrayals, she came within reach of the crown despite the absolute resistance of the men around her. Women like these did not win because they were exceptional in their abilities; they won because they refused to accept exclusion as inevitable.

Even in modern times, women who have broken through barriers have done so by force, not by patience. Angela Merkel was never expected to become the most powerful leader in Europe, yet she did so by outmaneuvering every opponent in her path, by playing the long game, by understanding that survival in power requires adaptability. Margaret Thatcher, regardless of one's views on her policies, did not rise to power because she was welcomed, she forced her way into a system that did not want her there, used its own structures against it, and remained in control for over a decade. These are not stories of compliance; they are stories of defiance, of understanding that power does not respond to politeness.

This is the lesson modern women must learn: power is not about worthiness. It is not about being the best, the brightest, the most prepared. It is about taking what is withheld. It is about recognizing that the game is not fair, and that fairness is not the goal. Those who rule do not care about fairness; they care about control. And control is won through strategy, through force, through an unrelenting willingness to demand what others are too afraid to claim. Women must stop waiting to be invited to the table. They must build their own, or they must take the one that exists and remake it in their image.

This does not mean abandoning integrity or morality; it means understanding that power is not a passive state. It must be maintained, defended, wielded with precision. Women who hesitate, who expect to be rewarded for their patience, will find themselves erased, replaced, forgotten. Those who act, who refuse to be managed, who recognize that opposition is not a reason to retreat but a sign they are doing something right, those are the women who will define the future. The world is not built to accommodate them. It is built to resist them. But resistance is not defeat. It is a sign that they are on the right path.

Women must be willing to fight. Not in the ways they are told to, not within the polite confines of what is deemed acceptable, but in the ways that actually shift power. They must fund each other, build networks that bypass traditional gatekeepers, refuse to be pacified by representation that does not come with real authority. They must study history, not the sanitized versions where women are footnotes, but the true stories of those who ruled, who conquered, who took what was denied to them. And they must understand that their enemies will never call them victorious. That is not how power works. The powerful do not announce their own downfall. They pretend nothing has changed, even as the ground shifts beneath them.

This is the reality of power: it will never be granted. It must be taken. And those who take it must be prepared for the consequences. They will be vilified. They will be underestimated. They will be told they are too much, too ambitious, too unwilling to play by the rules. But those who write the rules do so to keep others out. Women who understand this will stop waiting for change. They will become it.

Part 3

Power is an exercise in persistence, in seeing through the illusions that keep the majority controlled and contained while the few hoard wealth, influence, and authority. Women have been sold a story that patience leads to recognition, that competence ensures advancement, that playing by the rules guarantees a seat at the table. But history and reality both prove otherwise. Power is not granted on merit, it is taken by those who refuse to be denied. The moment women understand that, they stop waiting for a world that was never designed to include them. They stop being surprised when they are blocked, dismissed, or ignored. They stop wasting energy on the illusion of fairness and start channeling it into strategy.

The myth of gradual progress has been one of the most effective tools used to pacify women, keeping them hopeful rather than radicalized, patient rather than defiant. It is the idea that change is inevitable, that society naturally evolves toward greater inclusion, that if women just keep working hard, keep proving themselves, keep demonstrating their capability, eventually they will be rewarded. But change is not a natural occurrence. It is a battle, and the people who benefit from the existing power structures fight to prevent it. Every major gain in women's rights, whether political, economic, or social, was not granted by benevolent leaders who suddenly saw the light. It was forced into existence by women who refused to accept anything less.

Even the most basic rights, voting, education, property ownership, were won through disruption, through rebellion, through relentless opposition. The suffragists were not polite women who waited to be acknowledged; they were radicals who were beaten, imprisoned, force-fed, and demonized. The women who fought for workplace protections did not convince corporations to be kind; they shut down factories with strikes, they risked starvation to demand wages that allowed survival, they forced the system to change through direct action. And yet, the moment a gain is achieved, the illusion of progress is reasserted: "Look how far you've come. Now be grateful. Now stop." It is a pattern as old as power itself, small victories are used to prevent larger ones, and the moment women begin to expect more, they are told to be patient, to be satisfied, to stop demanding.

This is the cycle that must be broken. Women must recognize that power is not a negotiation, it is a struggle. And those who already hold it will never willingly give it up. The billionaire class does not maintain its wealth by accident; it is protected through policy, through propaganda, through a legal system that serves those who can buy influence. Political power is not neutral; it is shaped by laws that restrict access, by parties that function as gatekeepers, by networks that ensure the right kind of people remain in control. The corporate world is not a meritocracy; it is a fortress designed to allow the illusion of female advancement while ensuring that real decision-making remains in male hands.

Understanding this does not mean giving up, it means shifting strategy. Women must stop seeking permission. They must stop asking, hoping, waiting. They must recognize that power responds only to force, to disruption, to those who refuse to be managed. They must fund each other, build their own systems, bypass the structures that exist to exclude them. They must learn from history, not the history taught in schools, where women are footnotes, but the real history of those who took power when they were told they never could.

The most successful women in history did not expect to be treated fairly. They understood that power required both boldness and calculation, that survival depended not just on capability but on strategy. Cleopatra did not expect Rome to respect her authority, she secured her power through alliances, through manipulation, through sheer political acumen. Elizabeth I knew that marriage would make her a pawn, so she ruled alone, using every possible advantage to solidify her reign. Catherine the Great just out right seized the Russian throne by coup, knowing that waiting for legitimacy would mean her own downfall. These women were not given power, they took it, they held it, they refused to let go of it.

Even outside of monarchy, history is filled with women who understood that change required action, not patience. Harriet Tubman did not wait for slavery to end, she freed people herself, again and again, armed and unafraid. The women of the French Revolution did not simply support the movement; they demanded their own rights, stormed Versailles, led protests, and fought to be recognized. Women in labor movements, in civil rights struggles, in revolutions across the world have always been at the forefront, and yet their victories are often rewritten as footnotes, their contributions erased to uphold the myth that men shape history alone.

This erasure is intentional. If women knew how much power they had already seized throughout history, they would recognize their own strength. They would see through the lies that tell them they are powerless, that they must wait, that they must be satisfied with what they are given. And that recognition is dangerous to those who want to keep them contained.

The lesson of history is clear: those who succeed in power are those who refuse to wait for it. This is not about individual ambition; it is about collective strategy. Women must move as power moves, decisively, relentlessly, without hesitation. They must recognize that they will always be called too ambitious, too aggressive, too difficult, too much, and they must learn to embrace it. Because the alternative is obedience. And obedience is how power wins.

The women who will define the future are not those who seek approval, but those who demand authority. They are not those who ask politely, but those who take what is theirs. They are not those who hope for progress, but those who force it into existence. They do not wait. They act. And in doing so, they shift the balance of power, not by permission, but by force.

~2
Strategy

The Fine Line Between Influence and Subjugation

Power is never neutral. It is not something that exists in a vacuum, floating freely, waiting to be claimed by those who are best suited for it. Power is constructed, protected, hoarded, and wielded with precision. Those who have it defend it ruthlessly, and those who seek it are treated with suspicion, especially if they are women. The moment a woman displays ambition, the world reconfigures itself against her. She is no longer simply competent, intelligent, or strategic, she is cunning, manipulative, untrustworthy. Where men are celebrated for their ability to maneuver through power structures with skill and foresight, women are condemned for the same behavior. A man who makes calculated moves in politics or business is respected, admired for his pragmatism. A woman who does the same is accused of deceit, of playing games, of being cold, ruthless, unnatural. This is the double standard of power: the expectation that women, unlike men, must not only lead but do so in a way that is palatable, that does not upset the fragile sensibilities of those who have long controlled the game.

This expectation is one of the greatest obstacles modern women in politics and business face. They are not just fighting for access to power; they are fighting against the deeply ingrained cultural bias that tells the world they should not have it. A woman in politics must be twice as competent as her male counterparts just to be taken seriously, and even then, her every move is dissected for signs of illegitimacy. If she is too assertive, she is accused of being aggressive; if she is too measured, she is weak. If she negotiates, she is accused of scheming; if she does not, she is deemed ineffective. The contradictions are endless, the traps numerous, and the penalties severe. A woman in business faces the same scrutiny. She is expected to lead but not dominate, to succeed but not intimidate, to command but not control. The rules are clear: she must be strategic, but she must never be seen as strategic. Her power must appear accidental, an extension of her natural talent or charm, rather than the result of calculated effort.

History is full of women who understood these expectations and chose to defy them. They knew that waiting for the world to accept them was futile, that power would never come to them through goodwill or fairness. They took it instead, wielding strategy as their greatest weapon, navigating opposition with a level of precision that their male counterparts never had to master. Elizabeth I of England, Catherine the Great of Russia, and Cleopatra of Egypt were not merely rulers; they were architects of their own survival, each of them facing relentless hostility from the moment they stepped into power. They were framed as villains, their ambitions treated as threats, their intelligence viewed as dangerous. But they did not shrink under this weight. They understood that power is not about being liked, it is about being effective.

Elizabeth I inherited a throne fraught with instability, a nation divided by religious conflict, and a court that doubted her ability to rule without a husband. She was expected to marry, to cede her power to a man, to play the role of queen consort rather than sovereign. But she refused. She manipulated the expectations placed upon her, crafting an image of herself as the "Virgin Queen," married only to England, untouchable, ungovernable by any man. She used her femininity as a shield, wielding it strategically to avoid the fate of her mother, Anne Boleyn, and her cousin, Mary, Queen of Scots, both of whom were executed when they became too politically inconvenient. Elizabeth knew that to survive, she had to be both feared and adored, both commanding and elusive. She cultivated a persona that made her indispensable to her nation, ensuring that she could never be dismissed as just another woman on the throne.

Catherine the Great faced a different challenge. She was not born to rule; she was an outsider, a German princess married into the Russian imperial family, expected to produce heirs and remain in the background. But when her husband, Peter III, proved to be incompetent and widely despised, she saw her opportunity. She orchestrated a coup, rallying the military and nobility to her side, and took the throne for herself. She did not ask for legitimacy, she seized it. Her reign was marked by intelligence, reform, and expansion, but she was never forgiven for taking power in the first place. She was painted as a usurper, a seductress, a woman who schemed her way into a position she had no right to hold. Yet, despite the attempts to undermine her, she ruled for over three decades, strengthening Russia and solidifying her legacy as one of its greatest leaders.

Cleopatra's power was perhaps the most precarious of all. As a woman ruling in a world where female sovereignty was rare, she had to navigate not just internal threats but the growing dominance of Rome. She understood that survival required alliances, that to maintain her power, she had to be indispensable to those who had the ability to destroy her. She formed relationships with Julius Caesar and later Mark Antony, not out of romance but out of necessity, using every tool at her disposal to keep Egypt independent. Yet, history remembers her not as a brilliant strategist but as a seductress, as a woman who used beauty and manipulation rather than intellect and strategy. This distortion is no accident. It is part of the broader effort to discredit powerful women, to reduce their accomplishments to mere tricks of charm rather than acts of political genius.

These women's stories are not just relics of the past; they are blueprints for the women of today. The expectations may have changed in form, but the fundamental challenges remain the same. Women in power must still contend with the relentless scrutiny of their decisions, the need to justify every act of ambition, the requirement to be twice as good while receiving half the recognition. They must walk the fine line between influence and subjugation, ensuring that they are not seen as too calculating while knowing that without calculation, they will be overpowered.

The key to navigating this landscape is not to deny the reality of these double standards, but to use them to one's advantage. Women must learn the rules not to follow them, but to break them in ways that serve their interests. They must recognize that power is not about being liked or being fair, it is about being effective. This means understanding perception, controlling narratives, and refusing to be defined by the expectations imposed upon them. It means recognizing that strategy is not a dirty word, that ambition is not a flaw, and that influence is not something to be ashamed of. It means accepting that they will be called manipulative, ruthless, difficult, and refusing to be deterred by it.

For modern women in politics and business, the lesson is clear: do not wait for legitimacy to be granted. It never will be. Do not expect fairness; the game is not fair. Instead, study the structures of power, understand the forces at play, and move accordingly. Build networks that cannot be easily dismantled, craft images that make you indispensable, wield every resource available. Most importantly, do not apologize for doing what is necessary. The world will never fully embrace a woman in power, but that does not mean she cannot hold it. It only means she must take it knowing that opposition is inevitable, but irrelevance is optional.

Women have spent too long believing they must earn power in ways that men never have to. The truth is, power is not earned, it is taken. And those who take it must be prepared to be vilified. But they must also recognize that being feared, underestimated, or even hated is often a sign that they are doing something right. Because the only women who are never criticized are the ones who never challenged anything at all.

Part 2

Strategy is not a luxury for women who seek power; it is a necessity. Every move must be calculated, every decision weighed not just for its immediate effect but for its long-term consequences. A man can afford to act impulsively, to make mistakes, to rely on the assumption that his authority is inherent. A woman cannot. She must anticipate opposition before it arises, must counter accusations before they are made, must navigate a landscape designed to undermine her at every step. This is the fine line between influence and subjugation, the understanding that power is not just about having control, but about maintaining it in a world that resents her for holding it in the first place.

The expectation that women must be palatable in their pursuit of power is a carefully constructed form of control. It forces them to expend energy on appearance, on likability, on ensuring that their ambition does not make others uncomfortable. Men are granted the freedom to be ruthless, but a woman who exhibits the same qualities is deemed dangerous. This is not accidental. It is an attempt to keep women in a state of perpetual self-doubt, to make them question whether their pursuit of power is legitimate at all. The most effective way to neutralize a woman's power is to make her believe she must justify it.

The most powerful women in history understood that justification was a trap. They did not ask for permission to lead; they asserted their right to do so. Elizabeth I did not apologize for refusing to marry, she turned it into a political advantage, branding herself the "Virgin Queen" and making England her one true love. Catherine the Great did not wait for the Russian nobility to accept her, she took the throne by force and made them recognize her authority. Cleopatra did not shrink in the face of Rome's expansion, she used alliances, intelligence, and strategy to ensure Egypt remained a player on the world stage. These women were not loved by history. They were feared, they were slandered, they were framed as villains. But they ruled.

The modern woman who seeks power must embrace this lesson. She will be called manipulative simply for being strategic. She will be labeled unlikable for doing what men do without question. She must accept that she cannot control how she is perceived, only how she responds to that perception. The goal is not to be liked. The goal is to be effective.

For women in politics, the challenge is even greater. Every aspect of their existence is scrutinized in ways men never experience. Their clothing, their tone of voice, their family life, nothing is off-limits. The most powerful men in the world are allowed to be corrupt, incompetent, even outright criminal, and still retain their positions. A woman in politics, however, is expected to be flawless. If she makes a mistake, it is proof that women are unfit to lead. If she succeeds, she is accused of scheming. This double standard is not just frustrating; it is a weapon designed to keep women from even trying.

Hillary Clinton, Angela Merkel, Jacinda Ardern, each of these women faced intense scrutiny, often unrelated to their actual policies. Clinton was too ambitious, too robotic, too rehearsed. Merkel was too cold, too reserved, too uncharismatic. Ardern was too kind, too empathetic, too focused on consensus. The contradiction is clear: no matter how a woman leads, she will be criticized for it. The lesson? She must lead anyway.

Women in business face the same traps. They are expected to be collaborative, nurturing, approachable. If they are too authoritative, they are labeled difficult. If they are too accommodating, they are seen as weak. The myth of "empowerment" is often used as a distraction, companies love to highlight female executives while continuing to deny real systemic change. Women are encouraged to "lean in," to be more confident, to take their place at the table, while the structures that determine leadership remain designed to exclude them.

The key to navigating this is not to internalize these expectations, but to use them strategically. Women must stop trying to prove they deserve a seat at the table and start recognizing that the table was never built for them in the first place. The most successful women understand this. They do not waste time seeking validation from those who will never grant it. They build their own networks, cultivate their own sources of power, and ensure that they are too indispensable to ignore.

This does not mean abandoning strategy. On the contrary, it means mastering it. A woman in power must always think three steps ahead. She must anticipate how her words will be interpreted, how her actions will be framed, how her presence will be challenged. This is not paranoia; it is survival. Machiavelli wrote that a ruler must be both a lion and a fox, strong enough to command fear, but cunning enough to avoid traps. This is especially true for women, who must navigate a world that is more eager to see them fail than to see them lead.

The question then becomes: how does a woman wield influence without falling into the traps set for her? The answer is twofold. First, she must understand the systems she is working within. She must recognize that the rules were not written in her favor, and that playing by them will not ensure success. Second, she must be willing to defy expectations when necessary. A woman in power must choose her battles carefully, but when she strikes, she must do so decisively.

Cleopatra understood this well. She did not waste time trying to prove that she was as competent as a man. She demonstrated it through action. She controlled Egypt's wealth, maneuvered through Rome's political landscape, and ensured that her nation remained independent for as long as possible. Catherine the Great did not wait for legitimacy to be handed to her; she seized it, then rewrote the narrative of her rule to ensure she would be remembered as Russia's rightful leader. Elizabeth I did not try to convince her court that she was strong enough to rule alone; she simply ruled alone, making it clear that anyone who doubted her would regret it.

These women provide a model for modern leaders. They did not conform to expectations, they weaponized them. They understood that they would be villainized no matter what they did, so they made sure that if they were going to be painted as ruthless, they would at least be effective. They did not ask for approval. They took power and forced the world to reckon with them.

For women today, this means letting go of the need to be liked. It means understanding that no matter how they navigate power, there will always be those who resent them for it. It means refusing to waste time on those who expect them to be softer, gentler, more accommodating. The path to power is not paved with good intentions, it is carved through force of will, through strategy, through an unwavering commitment to the end goal.

The fine line between influence and subjugation is the awareness that a woman's power will always be contested. She must be willing to walk it without fear. She must understand that strategy is not manipulation, it is survival. She must accept that she will be criticized for the very qualities that make her successful. And most importantly, she must refuse to let that stop her. Power is not about being given a chance. It is about taking what others assume she cannot hold.

Part 3

Power is never taken without consequence. It is never seized without retaliation. The moment a woman asserts control, the moment she refuses to apologize for her ambition, the opposition sharpens its knives. This is the price of power, and every woman who has held it has paid it. The fine line between influence and subjugation is not just about knowing how to strategize; it is about understanding that the world will never forgive a woman for winning. A woman in power is tolerated only as long as she does not fully exercise that power. The moment she does, she is framed as dangerous, manipulative, unworthy. This is the game, and women must play it with eyes wide open.

The greatest mistake a woman in power can make is believing that if she follows the rules, she will be protected. The rules were never meant to serve her. They were meant to contain her. The illusion of meritocracy, the endless demand for women to prove themselves before they are taken seriously, these are all mechanisms of control. They are meant to keep women waiting, to make them hesitate, to prevent them from recognizing their own authority. This is why women must be ruthless in their understanding of power. They must know when to compromise and when to refuse. They must recognize when to play along and when to break the game entirely.

History has shown that the most successful women understood this. They were called villains, tyrants, manipulators, but they ruled. Elizabeth I, Catherine the Great, and Cleopatra did not waste time trying to convince the world of their worth. They demonstrated it. They did not attempt to soften themselves to make others comfortable. They understood that power does not accommodate, it dominates. These women did not rule because they were welcomed into power. They ruled because they refused to be denied it. They knew that influence was not enough; it had to be protected, reinforced, wielded with precision. And they knew that to hold power as a woman is to be in a constant state of battle.

For Elizabeth I, survival was the first priority. She inherited a throne that was fragile, a nation divided by religious conflict, and a court full of men who believed she was incapable of ruling alone. She played the game brilliantly, refusing to marry, refusing to be controlled, ensuring that she was always the most powerful figure in the room. She cultivated an image of divine authority, the "Virgin Queen" whose loyalty belonged only to England. This was not just branding; it was strategy. She understood that if she married, she would be reduced to a wife, her power transferred to her husband. She knew that her legitimacy as a ruler depended on her ability to remain singular, untethered to any man who could claim authority over her. Her reign was defined by this control, of her image, of her alliances, of her enemies. She did not seek approval; she commanded respect.

Catherine the Great took a different route. She was an outsider, a German princess never meant to rule Russia. But when her husband, Peter III, proved weak and incompetent, she saw her opportunity. She did not wait for the court's acceptance. She orchestrated a coup, rallied the military, and took the throne by force. She knew waiting for legitimacy would mean her downfall. She made herself indispensable, expanding Russia's empire, modernizing institutions, securing her place in history. But she was never forgiven for ambition. She was painted as a seductress, a usurper, a woman who stole power rather than inheriting it. This was the punishment for ruling as effectively as any man, her legacy distorted to fit the narrative that women must not desire power for its own sake.

Cleopatra faced an even greater challenge. She was a woman ruling in a world where female sovereignty was rare, contending with the rising power of Rome. She did not rely on brute force alone, she used intelligence, alliances, and negotiation. She knew Egypt's survival depended on her ability to navigate Rome's political landscape. She aligned with Julius Caesar, then Mark Antony, ensuring Egypt remained independent as long as possible. But history did not remember her as a ruler. It remembered her as a temptress. Her political acumen, strategic brilliance, ability to command loyalty, none of this was acknowledged. Instead, she was reduced to a woman who used beauty and manipulation to hold power. This is the ultimate distortion of female authority: when a woman is too powerful to be ignored, her legacy is rewritten to make her seem unworthy.

These historical lessons are not just stories; they are warnings. They reveal what happens when a woman takes power and refuses to surrender it. They show the patterns of backlash, the ways in which female authority is undermined, the relentless effort to discredit women who refuse to be controlled. And they offer guidance for modern women who face the same challenges. The question is not whether a woman will be criticized for her ambition. She will be. The question is whether she will allow that criticism to dictate her actions.

For modern women in leadership, whether in politics, business, or any other sphere, the most important lesson is to stop seeking validation from those who will never grant it. A woman who waits to be accepted will never reach the top. A woman who asks for permission will always be denied. Power is not something that can be earned through good behavior. It must be taken, wielded, defended. This means rejecting the expectation that women must be likable, that they must soften their edges, that they must make themselves small in order to be accepted. A man in power is allowed to be flawed, to be aggressive, to be decisive without apology. A woman must demand the same freedom.

This is not to say that strategy is unnecessary. On the contrary, it is everything. A woman who seeks power must be ten steps ahead, anticipating opposition before it arises, controlling narratives before they are written. She must recognize that perception is as powerful as reality, that the way she is seen will shape the way she is treated. This does not mean conforming to expectations. It means manipulating them. Elizabeth I used the image of the Virgin Queen to protect herself. Catherine the Great aligned herself with Russian nationalism to secure her reign. Cleopatra leveraged alliances to maintain Egypt's independence. Each of these women understood that power is not just about control, it is about perception.

For women today, this means refusing to play by rules designed to exclude them. It means refusing to be apologetic for ambition, authority, or taking up space. Criticism is inevitable, but irrelevance is optional. Women who seek power must be willing to be vilified, called difficult, labeled ruthless. Because the alternative is to remain in the background, overlooked, denied influence.

Power is not about being liked. It is about being effective. A woman who seeks to lead must be prepared for opposition, expect resistance, and understand that her success will always be framed as an anomaly rather than a rightful achievement. But she must lead anyway. She must take power anyway. She must recognize that the world does not reward patience, it rewards action. And those who hesitate, wait to be invited in, or believe fairness will one day prevail will find themselves waiting forever.

The fine line between influence and subjugation is the difference between those who accept power on dictated terms and those who take it on their own. Women must stop playing a game designed to keep them losing. They must redefine the rules, rewrite expectations, and refuse to be bound by limitations. The world does not make space for women in power. Women must take it. And once they do, they must never let it go.

~3
The Machiavellian Woman

Pragmatism Without Compromise

Power is not an exercise in morality; it is an exercise in control. Those who understand this succeed. Those who hesitate, who believe they can lead through sheer goodness, who expect fairness from a world built on domination, are swept aside. Machiavelli understood this reality better than most, and his lessons remain vital, not just for men seeking to maintain control, but for women who have been historically denied it. The Machiavellian woman is not a villain; she is a survivor. She is someone who recognizes that power is a game played with strategy, not sentiment. She does not apologize for ambition. She does not seek approval. She moves with precision, knowing that likability is a weapon used against her, that morality is often a luxury she cannot afford, and that sometimes, the only way forward is through ruthless pragmatism.

Machiavellianism is often framed as dishonesty, deception, even cruelty. But this is a misreading of *The Prince.* Machiavelli did not argue for evil; he argued for effectiveness. He understood that a leader must be feared more than loved, that perception shapes reality, that power must be defended at all costs. These principles are not just applicable to kings and statesmen, they are essential for women in leadership today. The myth that women must be liked to be successful is one of the most insidious traps of modern society. Men can be ruthless, decisive, and even cruel, and they are called strong. Women who do the same are branded as cold, calculating, untrustworthy. But history shows that the women who embraced Machiavellian pragmatism, the ones who understood when to fight, when to retreat, when to feign surrender, were the ones who survived, the ones who ruled, the ones who changed the world.

The demand for women to be likable is not a compliment; it is a form of control. It forces women to shrink themselves, to second-guess their authority, to manage not just their work but the fragile egos of those around them. A woman who is too assertive is "aggressive." A woman who is too reserved is "weak." A woman who speaks her mind is "difficult." A woman who plays by the rules and remains silent is "ineffective." No matter what she does, she is judged more for how she is perceived than for what she accomplishes. This is the impossible trap that keeps women in a permanent state of self-doubt, constantly adjusting, constantly accommodating, constantly trying to find the perfect balance between strength and approachability, a balance that does not exist.

Men are never held to this standard. A male CEO can fire employees without remorse, restructure companies with ruthless efficiency, and still be seen as a visionary. A female CEO who does the same is cruel, heartless, a disgrace to feminism. A male politician can wage wars, break alliances, lie to the public, and still be re-elected. A female politician who raises her voice too often is dismissed as hysterical. This is not an accident. It is the deliberate maintenance of a system that rewards men for power and punishes women for it. The women who have succeeded despite this reality, those who have truly held power, not just been granted symbolic positions, are the ones who understood that likability is a distraction. They did not waste time trying to be loved. They ensured they could not be ignored.

Elizabeth I ruled England for 45 years. She was not loved by all, but she was respected, feared, and ultimately, irreplaceable. She knew that likability was a liability. If she had played the role of the accommodating queen, she would have been married off, reduced to a political pawn. Instead, she crafted a persona that made her untouchable, strategically balancing between appearing untouchable and being undeniably powerful. She used flattery when necessary, threats when required, and deception when useful. She survived because she understood that survival was not about being liked, it was about being indispensable.

Catherine the Great came to power not by waiting, not by asking, but by taking. She orchestrated a coup against her own husband, secured the loyalty of the military, and transformed Russia into one of the most powerful empires of its time. She did not waste energy on whether her people adored her. She focused on making them need her. She knew that morality was often a privilege that rulers could not afford. She modernized Russia, expanded its territory, reformed its institutions, all while being vilified by those who resented her success. But she did not care. She had power, and she knew how to keep it.

Cleopatra was a master of perception, of political theater, of knowing exactly when to show strength and when to feign vulnerability. She did not rely on brute force; she relied on control. She understood that power is not just about winning battles, it is about ensuring that others believe you are always a step ahead. She formed alliances that kept Egypt independent for as long as possible, maneuvering through the treacherous landscape of Roman politics with skill and precision. And yet, she was reduced in history to nothing more than a seductress, a woman whose intellect was erased so that men could pretend she had no rightful claim to power. This is the ultimate lesson for women today: even when they are brilliant, even when they are strategic, even when they lead nations, they will still be framed as unworthy. The only response to this is to keep leading.

The Machiavellian woman does not waste energy fighting the perception of her. She uses it. She knows that people will always question her ambition, so she ensures that she is too useful to be discarded. She knows that she will be called manipulative, so she makes sure that every move she makes serves a purpose. She knows that she will be scrutinized more harshly than any man, so she leaves no room for mistakes. This is not unfair, it is reality. And reality is not something to complain about. It is something to master.

Tactical pragmatism is the ability to know when to fight, when to retreat, and when to feign surrender. Not every battle must be fought head-on. Not every victory must be declared immediately. The most powerful women in history understood this. They knew that sometimes, it is better to step back and let an enemy believe they have won, only to strike when they least expect it. They knew that patience is not weakness, that silence is not submission, that allowing an opponent to underestimate you is often the greatest advantage of all.

This is the difference between influence and true power. Influence is the ability to sway others, to convince, to persuade. Power is the ability to act without needing to ask for permission. The Machiavellian woman does not settle for influence alone, she ensures that she has power. She does not wait for approval. She does not seek validation. She moves with intent, understanding that the only thing that truly matters is control.

For women in leadership today, this means rejecting the demand to be likable. It means refusing to waste time proving their worth to those who will never accept them. It means knowing when to be quiet and when to be loud, when to concede and when to push forward. It means understanding that power is not about fairness, it is about leverage. And those who master leverage, those who recognize that likability is a myth designed to keep them docile, are the ones who will win.

The world does not reward kindness. It rewards strength. A woman who seeks power must be willing to be feared, to be underestimated, to be hated if necessary. Because the alternative is to remain powerless. And powerless is something the Machiavellian woman will never be.

Part 2

Power is not given to women. It is taken, wielded, and defended with a level of precision that men are rarely expected to master. Women who seek power must navigate a landscape designed to undermine them at every turn, a system that rewards their male counterparts for ambition while punishing them for the same trait. This is where the Machiavellian woman emerges, not as a villain, but as a strategist who understands that survival in leadership is not about playing fair, but about playing smart.

A woman who rises to power without understanding the ruthless pragmatism required to maintain it will be destroyed. This is not a warning, it is a fact, one that history has demonstrated time and time again. She will be questioned, scrutinized, and challenged at every turn. If she falters, if she hesitates, if she believes for even a moment that the world will be fair to her, she will be replaced. The expectation that women must be likable, that they must soften themselves to be palatable, is a carefully constructed trap meant to keep them from becoming too powerful. Men are allowed to be pragmatic, calculating, even cruel when necessary. Women who adopt these same tactics are labeled manipulative, deceitful, heartless. But power is not about morality, it is about effectiveness.

This is not to say that morality has no place in leadership. But morality alone does not win battles. It does not secure influence, does not protect against those who seek to undermine and dismantle. The Machiavellian woman understands this. She does not abandon morality, but she wields it strategically. She knows when to be kind and when to be ruthless, when to extend a hand and when to strike. She does not expect gratitude for her leadership, nor does she waste time defending her choices to those who will never support her. She moves forward, knowing that power demands action, not justification.

There is no clearer example of this than Elizabeth I. She understood that her authority as a female monarch would always be questioned, that her reign would be defined by the way she navigated the expectations placed upon her. If she had married, she would have been seen as weak, her power transferred to her husband. If she had ruled with too much force, she would have been deemed unfit, a woman incapable of controlling herself. So she walked the tightrope with precision, ensuring that she was never seen as too much of anything. She was firm, but not cruel. Charismatic, but not overly sentimental. She did not seek approval, she commanded respect. And she did it all while ensuring that no man could ever claim ownership over her authority.

Catherine the Great, too, was a master of pragmatic leadership. She did not rise to power through inheritance or goodwill, she took it. When her husband proved to be an incompetent ruler, she orchestrated a coup and seized the throne. But once she had power, she did not rule through sheer force alone. She used strategy, aligning herself with the Russian nobility, reforming laws to modernize the empire, and ensuring that those who opposed her had no choice but to accept her leadership. She understood that ruling as a woman required a different kind of strength. She could not afford to be indecisive, nor could she afford to rule with unchecked brutality. She balanced force with diplomacy, ensuring that even those who despised her recognized her necessity.

Cleopatra, perhaps more than any other historical figure, understood the art of political theater. She did not rule Egypt through military conquest; she ruled it through perception. She controlled her own narrative, knowing that her survival depended not just on her ability to govern, but on her ability to convince the most powerful men in the world that she was indispensable. She formed alliances with Julius Caesar and Mark Antony, not because she needed their approval, but because she understood that Egypt's independence depended on her ability to manipulate Rome's growing influence. And yet, despite her brilliance, history remembers her as a seductress rather than a strategist. This is the ultimate distortion of female power: when a woman rules too effectively, her accomplishments are rewritten as mere byproducts of her femininity, her intelligence erased in favor of a narrative that makes her power seem accidental rather than earned.

The lesson modern women must take from these rulers is not that they should abandon morality, but that they must understand when morality is a weakness and when it is a strength. Power is not maintained through idealism alone. It is maintained through control, of perception, of alliances, of every factor that determines success. A woman in leadership who expects to be rewarded for playing fair will find herself sidelined by those who do not. The Machiavellian woman does not seek to be loved. She seeks to be respected, feared if necessary, and above all, irreplaceable.

This does not mean abandoning ethics. It means recognizing that power is a game, and those who do not understand the rules will lose. A woman who enters politics expecting fairness will be eaten alive. A woman who starts a business believing that competence alone will carry her to success will be overlooked in favor of a less competent man who understands how to network, how to manipulate perception, how to demand what he has not yet earned. A woman who leads a movement without anticipating backlash, without understanding that power is something that must be defended as aggressively as it is acquired, will find her efforts undone by those who were willing to do what she was not.

Tactical pragmatism is not about deception; it is about knowing when to fight, when to retreat, and when to feign surrender. The most powerful women in history understood that not every battle was worth fighting. Sometimes, allowing an opponent to believe they have won is more advantageous than immediate victory. Sometimes, silence is more powerful than speaking. Sometimes, patience is the difference between survival and destruction. The Machiavellian woman knows this. She does not waste energy fighting battles she cannot win. She waits, she watches, she strikes when the time is right.

For modern women, this means learning to recognize the forces working against them before those forces strike. It means anticipating opposition, understanding that every decision will be questioned more harshly than a man's, and refusing to be derailed by it. It means learning how to navigate a world that was not designed for them without being consumed by it. It means understanding that power is not just about gaining control, but about keeping it.

The demand for women to be likable is the most insidious trap of all. It forces them to waste time proving that they are not too ambitious, not too aggressive, not too threatening. It keeps them distracted, apologizing for their success, explaining their choices, justifying their authority. The Machiavellian woman rejects this entirely. She does not explain. She acts.

Likability is a myth designed to keep women out of power. The sooner women realize this, the sooner they can stop seeking approval and start demanding control. Because the world does not reward kindness. It rewards strength. And those who understand this, those who stop waiting to be accepted and start taking what is theirs, are the ones who will define the future.

Part 3

The Machiavellian woman is a figure that history refuses to embrace yet cannot erase. She is the woman who did not ask for power but took it, who refused to wait for permission but instead created her own authority, who understood that fairness was a myth and that power rewards those willing to wield it without apology. She is vilified, misunderstood, framed as dangerous not because she was cruel, but because she was effective. The expectation that women must be liked, that they must temper their ambition, that they must make their power palatable to a world that resents it, this is the ultimate tool of subjugation. A woman who seeks power must understand this trap and refuse to fall into it. She must not waste time defending herself against accusations of ruthlessness. She must not soften her authority to appear more acceptable. She must not explain away her success as luck, timing, or the generosity of others. The world does not ask men to apologize for their ambition; it does not demand that they justify their authority. The Machiavellian woman rejects the expectation that she must be anything other than what is necessary to win.

The greatest weapon used against women in power is perception. A woman who is decisive is called controlling. A woman who is pragmatic is called manipulative. A woman who strategizes is called deceitful. These labels are designed to disarm, to keep women doubting themselves, to make them hesitate at the very moment they should act. The Machiavellian woman knows this. She does not fight against these perceptions; she uses them. She understands that if she is going to be framed as dangerous, she might as well make herself too dangerous to be ignored. She does not waste energy trying to be liked. She ensures that she is feared, respected, and above all, irreplaceable.

Power is not about being the best. If it were, the most competent, intelligent, and capable people would be in charge. But history shows that power is not a meritocracy. It is about perception, control, and the ability to outmaneuver those who seek to destroy you. A woman who seeks power must be prepared for betrayal, for sabotage, for attacks that have nothing to do with her abilities and everything to do with her existence. The Machiavellian woman does not expect loyalty, she secures it through necessity. She does not expect fairness, she creates her own advantage. She does not rely on the system to recognize her worth, she ensures that the system cannot function without her.

This is what Elizabeth I understood. She knew that her legitimacy as a ruler would always be questioned, that she would always be compared unfavorably to the kings who came before her. She knew that if she married, her power would be diminished, so she never did. She cultivated an image of divine authority, ensuring that no one could claim ownership over her. She did not trust blindly, she surrounded herself with advisors who were useful, but she never allowed them to believe they were indispensable. She was careful with her words, selective with her allies, ruthless with her enemies. She knew that power was not about fairness, it was about survival.

Catherine the Great took this lesson even further. She did not wait for the court to accept her. She orchestrated a coup, took the throne by force, and made it clear that she would not be a passive ruler. She did not rule through brute strength alone, she ruled through intelligence, diplomacy, and the careful cultivation of loyalty. She understood that in order to maintain power, she had to be useful to those around her. She modernized Russia, expanded its empire, ensured that her reign was one of undeniable success. And yet, history still painted her as a woman who had stolen power rather than earned it, because the world cannot stand a woman who rules without apology.

Cleopatra, perhaps more than any other ruler, understood the importance of perception. She did not rule through military conquest; she ruled through political mastery. She knew that power was about alliances, about making herself indispensable to those who could destroy her. She aligned herself with Julius Caesar, then with Mark Antony, not out of weakness but out of strategy. She made herself too valuable to be discarded, ensured that Egypt remained independent for as long as possible. But history did not remember her as a ruler, it remembered her as a seductress. Her intelligence, her strategic brilliance, her ability to navigate the most dangerous political landscape of her time, all of it was erased in favor of a narrative that made her power seem accidental rather than earned.

The Machiavellian woman does not wait to be understood. She does not seek validation from a world that will always question her authority. She moves forward, knowing that she will be judged unfairly, knowing that she will be criticized more harshly than any man, knowing that she will be seen as a threat. And she does not let that stop her. Because power is not about being accepted, it is about being in control.

This is the lesson for women in leadership today. They will be judged whether they act or remain silent, whether they lead or follow, whether they fight or submit. There is no path to power that is free of opposition. The only choice is whether they will let that opposition define them or whether they will move past it. A woman who seeks power must be willing to be vilified. She must be willing to be feared. She must be willing to be called ruthless, difficult, unlikable. And she must never let those words weaken her resolve.

Tactical pragmatism is the ability to know when to fight, when to retreat, and when to feign surrender. It is the art of playing the long game, of knowing that not every battle must be won immediately, that sometimes the best move is to wait until the right opportunity presents itself. The Machiavellian woman does not fight every fight, she fights the ones that matter. She does not waste energy proving herself to those who will never accept her, she focuses on consolidating her power so that their acceptance is irrelevant. She does not let emotion dictate her decisions, she calculates, she strategizes, she ensures that every move she makes is one that strengthens her position.

This is not about abandoning ethics. It is about understanding that morality alone does not win wars, secure leadership, or protect those who refuse to engage in necessary battles. A woman who seeks power must understand that she will be judged by different standards, expected to be both strong and soft, decisive and accommodating, powerful and humble. She must reject these contradictions entirely. She must refuse to shrink herself to fit expectations of those who do not want her to succeed.

For modern women, this means anticipating resistance before it arises. It means knowing that every action will be scrutinized more harshly than a man's and refusing to let that scrutiny dictate decisions. It means recognizing that power is not about waiting for permission, it is about taking what others assume they cannot hold. The demand for women to be likable is not just unrealistic; it is a tool of control. It keeps women focused on pleasing rather than leading, making others comfortable rather than securing their own success. The Machiavellian woman understands that likability is a distraction. She does not waste time trying to be accepted. She ensures she is necessary.

The world does not reward kindness. It rewards strength. And those who understand this, those who stop apologizing, waiting, seeking approval, are the ones who will define the future. The Machiavellian woman does not hope for fairness. She does not expect justice. She does not wait for an invitation. She takes power, knowing she will never be welcomed, and holds it, knowing the world will never forgive her for it. But she does not care. Because she is not here to be forgiven. She is here to win.

~4

Enemies & Obstacles

The Reality of Opposition

Opposition is not an accident. It is not a rare misfortune that befalls only a few women who rise too high, who speak too boldly, who refuse to be manageable. It is inevitable, built into the very foundation of a world that has long been ruled by those who see female power as an existential threat. A woman who seeks power must not only expect opposition; she must assume it will come from every direction, subtle and overt, institutional and personal, calculated and reactionary. It is the price of refusing to stay in place. The greater the ambition, the greater the resistance. The higher the climb, the more violent the efforts to pull her back down. This is not a reason to retreat. It is a sign that she is doing something right.

The world does not fear powerless women. It does not sabotage them, ridicule them, attempt to contain them. It does not need to. The powerless are already controlled. But a woman who rises, who asserts herself in politics, in business, in any sphere where she is not meant to lead, immediately attracts opposition. The system recognizes the threat before she even has a chance to wield it. She is told she is unlikable, unelectable, unfeminine. Her qualifications are questioned. Her ambition is framed as dangerous. If she is assertive, she is aggressive. If she is reserved, she is weak. If she succeeds, she is accused of being calculating. If she fails, she is held up as proof that women are unfit to lead. She cannot win by playing fair. And so, she must stop trying to.

Today's patriarchal power structures, corporate, religious, and political, are not just resistant to female leadership; they are structured to prevent it. The corporations that put women on magazine covers to celebrate their "firsts" are the same ones that still pay them less than men, that still deny them venture capital, that still refuse to make space for them at the highest levels. The churches that celebrate virtuous, obedient women are the same ones that have spent centuries barring them from the pulpit, writing doctrine that ensures they remain subordinate. The politicians who claim to support gender equality are the same ones who vote against policies that would make it a reality. These structures are not built to include women in leadership. They are built to prevent them from obtaining real power while giving the illusion of progress.

The illusion is important. It is what keeps women believing that if they just work hard enough, if they prove themselves enough, if they are patient enough, they will eventually be allowed in. It keeps them striving within a system that has already determined they will not be permitted to reach the top. It ensures that they waste time on likability rather than strategy, on proving themselves rather than taking what is theirs. It convinces them that the backlash they face is their fault, that if they had just been a little softer, a little less ambitious, a little less threatening, they would not have been met with resistance. This is the greatest lie of all. Resistance does not happen because a woman is too much. It happens because she is powerful, and the world has never been comfortable with powerful women.

History offers proof of this in every era, in every culture, in every power structure. No woman who has ever ruled has done so without being framed as unnatural, untrustworthy, dangerous. They have been branded as witches, as seductresses, as traitors. They have been erased, rewritten, reduced to cautionary tales. Catherine the Great was one of the most successful rulers in Russian history, yet she was slandered relentlessly, her intelligence overshadowed by lurid rumors. Cleopatra was one of the most brilliant political minds of her time, yet history remembers her as a woman who used seduction rather than strategy. Elizabeth I ruled England for nearly half a century, navigating war, religious conflict, and political betrayal, yet she was constantly forced to prove that she was as strong as any king. These women did not face opposition because they were unqualified. They faced opposition because they refused to be ruled.

The strategies these women used to turn opposition into opportunity are still relevant today. They understood that power is not about fairness. It is about leverage. They built alliances where they needed to, cut ties when necessary, refused to be sentimental about those who sought to undermine them. They understood that betrayal was inevitable and prepared for it rather than allowing themselves to be shocked by it. They knew that backlash was not a sign of failure, but of success, that the more fiercely they were resisted, the more effective their leadership had become.

A modern woman who rises must do the same. She must anticipate that opposition will come from both expected and unexpected places. She must recognize that not all resistance will be direct; some will come in the form of subtle sabotage, exclusion from key conversations, shifting goalposts that make it impossible for her to succeed in the way men do. She must be willing to be strategic, to see betrayal before it happens, to recognize who is truly an ally and who is merely waiting for the right moment to push her aside. She must not let herself be distracted by likability politics, by the expectation that she should soften herself to be accepted. She must know that if she is not experiencing resistance, she is not pushing hard enough.

The most important lesson is this: power does not tolerate women who refuse to be controlled. A woman in leadership will never be allowed to operate under the same rules as a man. She will be tested in ways that he will never have to endure. She will be questioned more, challenged more, undermined more. She will be expected to justify her presence in rooms where men are simply assumed to belong. She will be warned, threatened, attacked. She will be told she is not wanted. But she must lead anyway.

To prepare for this, a woman must build her own structures of power. She must not rely on the systems that were designed to exclude her. She must not depend on being granted legitimacy by those who do not want her to have it. She must cultivate her own networks, her own influence, her own authority. She must ensure that when backlash comes, she is too entrenched to be removed easily. She must learn from the women before her, how they maintained control, how they adapted, how they survived in a world that wanted them gone.

And she must understand that opposition is not proof that she is failing. It is proof that she is a threat. This is the most powerful position she can be in. A woman who is not feared is a woman who is not yet in control. The backlash, the resistance, the attempts to discredit her, these are signs that she has stepped into real power. And she must never step back.

For centuries, women have been told that if they just wait, if they just prove themselves, if they just play by the rules, they will be rewarded. But the women who actually ruled, who actually shaped history, knew that this was a lie. They did not wait. They did not ask. They took. They fought. They endured. And they won. The woman who seeks power today must be prepared to do the same. She must expect resistance. She must welcome it. Because opposition is not an obstacle. It is confirmation that she is on the right path.

Part 2

Opposition is not just something women in power face, it is the defining feature of their journey. The world does not merely discourage female ambition; it actively works to punish it. Those who rise are met with resistance not because they are unqualified or unworthy, but because their very presence disrupts the structures built to exclude them. Every woman who has ever sought to lead, to govern, to command, has encountered systemic obstacles designed to slow her, silence her, or force her into submission. These barriers are not accidents. They are the architecture of a patriarchal order that has long understood that keeping women out of power requires more than just denying them access, it requires exhausting them, making them question themselves, forcing them to fight battles that men never have to fight.

Women in positions of power do not merely lead; they battle for the right to lead. Every day. Every decision. Every moment. It is a war fought on multiple fronts, within the corporate boardroom, the political arena, the cultural landscape, and even within the very movements that claim to support gender equality. A woman's authority is never assumed; it is always questioned. Her success is never credited solely to her talent; it is attributed to luck, to manipulation, to the generosity of others. Her mistakes are not treated as human errors, but as proof that women should never have been in power to begin with. She is never simply a leader, she is a *female* leader, as if the modifier is necessary, as if it sets her apart, as if it diminishes her legitimacy.

The structures that resist female leadership are not disorganized or accidental. They are deeply entrenched, intentionally maintained, and constantly evolving to preserve male dominance.

The corporate world loves to tell stories of progress. Every time a woman is promoted to a high-ranking position, she is held up as proof that things are changing. But the numbers do not lie. Women still occupy only a fraction of executive leadership positions. They are still underpaid, underfunded, and underrepresented in the highest levels of decision-making. The celebrated "firsts" are meant to distract from the fact that systemic inequality remains intact. The glass ceiling is not simply an obstacle to be broken, it is a mechanism of control. It keeps women striving within a system that is rigged against them, convincing them that if they just work hard enough, if they just prove themselves a little more, they will finally be allowed in.

The reality is far harsher. Women in corporate leadership face not only bias but active sabotage. Their decisions are scrutinized more heavily than their male counterparts'. They are expected to mentor, to nurture, to take on additional emotional labor, all while being judged as too soft if they show empathy and too harsh if they enforce standards. If they negotiate for higher salaries, they are labeled as difficult. If they do not, they are seen as weak. They are held to impossible standards, constantly balancing between authority and likability, an equilibrium that does not exist.

The companies that publicly celebrate diversity initiatives are the same ones that systematically exclude women from the real centers of power. They create diversity committees, publish equality statements, and highlight token female executives, all while maintaining the same structures that prevent women from truly leading. When a woman does make it to the top, she is often placed in precarious positions, given roles where failure is likely, where resources are limited, where she can be easily blamed if things go wrong. And when she falls, she is treated as evidence that women are not fit to lead.

The world's major religious institutions have long been the guardians of patriarchy, ensuring that women remain subordinate, silent, and controlled. They write doctrine that frames female leadership as unnatural, that ties women's worth to obedience, that makes defiance a sin. They claim moral authority while institutionalizing female inferiority, barring women from priesthoods, from leadership, from interpreting the sacred texts that have been weaponized against them.

Even in modern religious movements that claim to support gender equality, the resistance to female leadership is clear. Women may be allowed to preach, but only under certain conditions. They may hold titles, but only if they conform to traditional expectations. The moment a woman asserts full autonomy, the backlash is swift. The moment she demands the same authority as a man, she is deemed heretical, radical, dangerous.

This is not simply about theology, it is about control. Women who are taught to submit in the name of religion are more likely to submit in other areas of life. They are less likely to challenge male authority, less likely to seek power, less likely to believe they deserve leadership. Religious institutions do not just resist female leadership within their own walls; they shape the cultural attitudes that keep women out of power in every sector of society.

Politics is one of the most hostile arenas for women. The attacks are relentless, the scrutiny invasive, the expectations suffocating. A male politician can be mediocre, corrupt, even incompetent, and still maintain his position. A female politician must be extraordinary just to be considered legitimate. And even then, she will face attacks not just on her policies but on her very existence. She will be questioned about her family, her appearance, her personality. Her tone will be dissected. Her decisions will be held to impossible standards. If she compromises, she is weak. If she refuses, she is inflexible. If she succeeds, she is lucky. If she fails, she is proof that women should not be in power.

The political establishment does not simply resist women, it humiliates them, grinds them down, exhausts them until they either conform or quit. Women who run for office receive more threats, face more harassment, and are more likely to be criticized for their personal lives than men. Their ambitions are seen as unnatural, their authority as suspect. They are told they are unelectable, even when they win. They are accused of being too emotional, even when they remain composed. They are expected to prove themselves endlessly, while men are allowed to fail upward.

The strategies of women who have survived these institutions are not about winning a fair fight. They are about understanding that fairness is an illusion. They are about knowing that power does not accommodate, it is seized. The most powerful women in history did not wait for permission. They did not waste time proving they were good enough for systems designed to exclude them. They took what they needed. They built their own networks. They outmaneuvered those who sought to destroy them.

Elizabeth I did not waste time convincing her court that she was fit to rule. She declared herself divinely ordained and made it clear that questioning her authority was treason. Catherine the Great did not plead for legitimacy, she staged a coup and rewrote Russian history in her own image. Cleopatra did not ask Rome for permission to lead Egypt, she ensured that Egypt was too valuable to be ignored.

These women understood that opposition is not a sign of failure, it is a sign of power. The more fiercely they were resisted, the more dangerous they had become. They did not shrink. They did not soften. They did not beg to be accepted. They ruled. A modern woman who seeks power must do the same. She must anticipate betrayal before it happens. She must build alliances that cannot be easily dismantled. She must understand that she will be attacked, questioned, discredited, but she must not let it stop her. She must be willing to be hated, because the alternative is to be irrelevant.

She must recognize that the world will not welcome her ambition. It will call her difficult. It will call her ruthless. It will call her dangerous. And she must embrace it. Because opposition is not a sign that she is failing, it is proof that she is winning. A woman who does not face resistance is a woman who has not yet threatened the system. A woman who does not encounter backlash is a woman who has not yet taken real power. And a woman who understands this, who sees opposition not as a barrier but as confirmation that she is on the right path, is the woman who will lead. Not by permission. Not by approval. But by force of will.

Part 3

A woman who rises to power will never be allowed to exist in peace. She will be questioned, challenged, undermined, and attacked from all sides, not because she is weak or unqualified, but because her very existence as a leader disrupts the established order. The backlash she faces is not incidental. It is a strategy, a coordinated effort to exhaust her, distract her, and ultimately remove her. The moment she understands that opposition is not something to be feared but expected, she shifts from being a target to being a strategist.

The most effective women in history have never viewed resistance as an obstacle; they have viewed it as confirmation that they are succeeding. Every act of sabotage, every attempt to discredit them, every whisper campaign against them only proves that they have power worth fearing. The difference between those who rise and those who fall is not the presence of opposition, it is the ability to withstand it.

Opposition does not always come in the form of direct confrontation. Often, it is more insidious. It is the slow, steady erosion of a woman's credibility, the subtle discrediting of her authority, the endless cycle of scrutiny designed to wear her down until she either gives up or becomes so consumed with fighting back that she cannot focus on leading. The first and most effective weapon used against a woman in power is the credibility trap. No matter how qualified she is, she will always be forced to prove herself in ways that men are not. She will be asked to justify every decision, every appointment, every policy, every success. If she falters, it will be seen as evidence that women are unfit to lead. If she succeeds, the credit will be given to someone else, to circumstance, to luck, to a male mentor who "helped" her get there.

Likability is another powerful tool used to keep women in check. A male leader is allowed to be aggressive, blunt, even tyrannical, and his style is praised as strong and effective. A female leader who exhibits the same behavior is labeled difficult, demanding, cold. The pressure to be likable forces women to dilute their authority, to temper their words, to make themselves smaller in order to avoid backlash. But the paradox is that no matter how likable a woman tries to be, it will never be enough. If she is too warm, she is seen as weak. If she is too firm, she is called unapproachable. If she is too charismatic, she is accused of being manipulative. If she lacks charisma, she is dismissed as uninspiring. The demand for likability is not a measure of leadership, it is a method of control.

Another common tactic is the loyalty test. Women in leadership are often expected to be responsible for the success of other women, held to an impossible standard in which they must lift up everyone while also proving that they are not playing favorites. When a male leader surrounds himself with male advisors, it is seen as normal. When a woman does the same with female allies, it is seen as suspicious, evidence that she is practicing gender-based favoritism. And yet, if she does not support other women, she is accused of betraying them. There is no correct answer, only a constant state of scrutiny designed to keep her distracted and defensive.

The most brutal weapon used against powerful women is character assassination. Men in power can be corrupt, incompetent, even criminal, and they will still be able to hold onto their authority. A woman's mistakes, however small, will be magnified, scrutinized, and used as evidence that she is unfit to lead. If there are no real mistakes to exploit, they will be invented. Rumors will be spread, false narratives will be built, scandals will be fabricated. The goal is not just to remove her from power but to ensure that she is permanently discredited, that her reputation is destroyed so thoroughly that no one will ever see her as a leader again.

The women who have survived these attacks are those who understood that power is not about fairness. They did not waste time trying to defend themselves against every accusation, did not allow themselves to be drawn into battles that would only serve to drain their energy. They knew when to respond and when to ignore. They built networks of power that could not be easily dismantled, surrounded themselves with people who were loyal not out of obligation but out of shared interest. They prepared for betrayal, anticipated sabotage, and made sure that when the attack came, they had already positioned themselves in a way that made them impossible to remove.

Elizabeth I survived because she refused to play by the rules that had been designed to destroy her. She knew that if she married, she would become a pawn, so she remained single, controlling her image so effectively that she turned her unmarried status into a symbol of strength rather than weakness. Catherine the Great knew that her legitimacy would always be questioned, so she aligned herself with Russian nationalism, making it impossible for her enemies to remove her without destabilizing the empire. Cleopatra understood that perception was as powerful as reality, so she crafted an image that made her indispensable to Rome, ensuring that her alliances with Julius Caesar and Mark Antony were not seen as submission, but as strategic partnerships.

For women today, the lessons remain the same. The system is designed to test them, to wear them down, to force them to prove themselves in ways that men never have to. The only way to survive is to refuse to play by those rules. A woman in power must be willing to be hated. She must accept that she will never be fully embraced, that she will always be viewed with suspicion, that there will always be those who seek to destroy her. She must not waste time trying to change their minds. She must focus on consolidating power, on building alliances that cannot be easily severed, on making herself too necessary to be removed.

Opposition is inevitable. Betrayal is inevitable. Resistance is inevitable. But none of these things are reasons to stop. They are proof that a woman has become dangerous, that she has entered a space where her power is real. The world does not fight powerless women. It fights those who threaten to change it. And that is the greatest confirmation that a woman is on the right path.

The only way to defeat an enemy is to stop fearing them. Women who seek power must understand that backlash is not a sign of failure, it is the cost of success. They must be prepared for the fight, not as victims, but as rulers who have already decided that they will not be moved. A woman who is ready for opposition is a woman who cannot be broken. And a woman who cannot be broken is a woman who will win.

~5
The Woman Prince

Ruling in a World That Fears You

A woman who rules is never just a leader. She is a spectacle, an anomaly, a target. The world does not simply watch her, it scrutinizes her, waiting for any misstep, any crack in her armor, any moment of perceived weakness to use against her. A male leader can be corrupt, incompetent, even reckless, and still retain his authority. A woman in power has no such leeway. She must be flawless, yet she must not appear too perfect. She must be decisive, yet she must not be threatening. She must be authoritative, yet she must not appear domineering. Every move she makes is judged against a standard that does not exist for men, and the contradictions are deliberate. They are meant to exhaust her, to keep her second-guessing herself, to ensure that she is always balancing on an impossible tightrope.

No woman who has ever ruled has been allowed to lead without resistance. From Cleopatra to Catherine the Great, from Elizabeth I to modern-day female heads of state and CEOs, the burden of expectation placed on women in power is one that few men could ever survive. They are expected to be both warriors and nurturers, to lead with strength but also with warmth, to make difficult decisions while ensuring that they do not upset the fragile egos of those around them. If they are too tough, they are labeled cruel, heartless, unlikable. If they show emotion, they are weak, hysterical, irrational. These standards are not meant to be fair. They are meant to make women question themselves, to make them hesitant, to make them easier to remove.

The first lesson a woman in power must learn is that she will never be fully accepted. No matter how skilled she is, no matter how many battles she wins, no matter how much she achieves, there will always be those who resent her presence. She must stop expecting fairness. She must stop trying to win people over who will never respect her. She must accept that her authority will always be questioned and prepare accordingly. A man in power is given the benefit of the doubt. A woman must create an environment where doubt is not an option.

This begins with the psychology of leadership. Machiavelli wrote that it is better to be feared than loved, but for a woman, the calculation is more complex. If she is only feared, she will be despised, undermined, and eventually removed. If she is only loved, she will be dismissed as ineffective, controlled by those around her, unable to make difficult decisions. The woman prince must strike a balance between fear and respect. She must make it clear that she is not to be challenged, that disloyalty will not be tolerated, that she is not here to be anyone's friend. But she must also be careful not to create unnecessary enemies. She must be strategic in her punishments, calculated in her rewards, ensuring that those who follow her do so not just out of obligation, but out of recognition that she is the only one capable of leading.

Elizabeth I mastered this balance. She ruled a divided nation, surrounded by those who believed that a woman could never be a true sovereign. She played the role of the benevolent mother of England, ensuring that she was beloved by the people, while also making it clear that she would not hesitate to execute anyone who posed a threat to her rule. She did not allow herself to be softened by the expectations placed upon her. She knew that if she faltered, if she showed even a moment of indecision, her enemies would take advantage. She was feared, but she was also respected. She understood that to rule as a woman, she could not simply be a monarch, she had to be a symbol, an untouchable force that could not be easily replaced.

Catherine the Great took a different approach. She knew that she could never fully erase the stigma of having taken the throne through a coup, so she made herself indispensable to Russia. She ruled not through brute force, but through sheer effectiveness. She modernized the empire, strengthened the military, expanded the economy. She did not waste time trying to be loved, but she also did not make enemies carelessly. She ensured that those who benefited from her rule had a reason to support her. She did not demand loyalty, she made it clear that her leadership was the best option available.

Cleopatra understood the power of perception. She ruled in a time when a female pharaoh was an anomaly, an exception rather than the rule. She knew that to maintain her position, she had to control how she was seen. She crafted an image of divine authority, making herself inseparable from the idea of Egypt itself. She formed alliances strategically, aligning with powerful Roman leaders not out of submission, but out of necessity. She understood that leadership was not just about ruling, it was about ensuring that others believed she was the only one capable of doing so.

The modern woman in power faces a different battlefield, but the tactics remain the same. She must navigate a world that still resists her authority, that still sees her as unnatural, that still demands that she justify her leadership in ways that men never have to. She must be ready for isolation. The higher she climbs, the fewer true allies she will have. She must be prepared for betrayal. The more successful she is, the more those around her will seek to bring her down. She must understand that criticism is inevitable, that no decision she makes will ever satisfy everyone, that no victory will ever silence all of her detractors.

She must learn to rule without apology. She must stop seeking approval from those who will never give it to her. She must recognize that power is not about being liked, it is about being in control. A woman who waits for permission to lead will never lead. A woman who seeks to prove herself to those who doubt her will waste time that could be spent consolidating her authority. The woman prince does not ask to rule. She rules.

This is not to say that a woman in power must be cruel. She must be strategic. She must understand when to show strength and when to show restraint, when to punish and when to forgive, when to stand firm and when to adapt. She must ensure that those around her do not see her as easily replaceable. She must make herself necessary. She must learn to command respect not by demanding it, but by making it clear that there is no alternative to her leadership.

The burden of expectation placed upon women in power is suffocating, but it is also revealing. It shows who is truly fit to lead. A woman who can navigate the endless scrutiny, who can withstand the constant resistance, who can rise despite the forces that seek to pull her down, is a woman who has already proven herself far stronger than any man who has been handed power without question. The world does not fear weak women. It fears those who refuse to be controlled.

The woman prince is not concerned with being understood. She is not concerned with being accepted. She is concerned with ruling. She is concerned with ensuring that her power is real, that it is defended, that it is permanent. She does not waste time explaining herself. She does not soften herself to make others comfortable. She does not apologize for her ambition.

The world will never fully embrace a woman in power. But a woman who understands this, who does not expect acceptance, who does not waste energy seeking validation, is a woman who will not be stopped. She does not rule to please others. She rules because she has decided that she will not be moved. And in the end, that is what makes her unstoppable.

A woman who rules does so in a world that was never built to accommodate her. She is not simply a leader; she is a disruption, an aberration, a figure who must constantly justify her authority in ways men never have to. The standards set for her are impossibly high, and they are designed not to be met, but to exhaust her, to keep her occupied with proving her legitimacy instead of wielding her power. A man in power is allowed to be flawed, to fail, to make mistakes and still recover. A woman in power walks a different path, one where any misstep can be used to destroy her, where her successes are treated as exceptions rather than the rule, where she must always be better, sharper, and more strategic just to hold on to what men inherit without question.

Women who seek power must understand that they are playing a different game with different rules, and those rules were not written in their favor. They will be scrutinized relentlessly, their actions judged not on their effectiveness but on their adherence to an arbitrary and shifting standard of acceptability. They will be expected to be strong but not threatening, confident but not arrogant, authoritative but not harsh. They must command respect without demanding it too forcefully, they must navigate opposition without appearing combative, they must balance their ambition so carefully that it does not unsettle those who resent their presence. It is an impossible balance, designed to keep them in a perpetual state of self-doubt, to force them to waste time adjusting rather than leading.

This is why the most powerful women in history have not wasted time trying to meet these expectations. They have rewritten the rules entirely. They have understood that their power does not depend on being liked or accepted, but on being necessary, indispensable, impossible to remove. They have ruled with a combination of fear and respect, knowing that to be loved without fear is to be dismissed, and to be feared without respect is to be overthrown. They have walked the fine line between inspiring loyalty and demanding obedience, between appearing accessible and remaining untouchable. And they have done it in a world that would rather see them fail than acknowledge that they were always more capable than the men who sought to replace them.

To understand the psychology of leadership as a woman is to understand that perception is as powerful as reality. A woman who appears uncertain will be treated as weak. A woman who appears too confident will be treated as arrogant. A woman who wields power too forcefully will be resented, but a woman who wields it too cautiously will be ignored. The way she presents herself, the way she carries her authority, the way she controls the room matters as much as the decisions she makes. Elizabeth I did not rule England for nearly half a century by accident, she ruled by crafting an image that made her untouchable. She understood that to survive, she could not simply be a queen; she had to be more than that. She had to be a myth, an idea, a figure that transcended ordinary power struggles. She had to make it so that any attack on her was an attack on England itself.

Catherine the Great did the same. She did not inherit power, she took it. She understood that she would always be seen as an outsider, as a woman who had no rightful claim to rule Russia. So she made herself indispensable. She expanded the empire, modernized its institutions, strengthened its economy, and ensured that even those who despised her had no choice but to accept that she was the best ruler they could have. She did not concern herself with whether she was liked. She concerned herself with making it impossible for anyone to remove her without destroying the nation in the process.

Cleopatra understood the power of alliances and perception better than anyone. She did not try to prove to her enemies that she deserved the throne. She ensured that Egypt remained strong by forging strategic relationships with Rome, knowing that her survival depended on her ability to navigate the ambitions of powerful men. She did not rule by brute force, she ruled by making herself essential. She controlled the narrative of her reign, presenting herself not just as a queen, but as a living goddess, a figure whose authority was unquestionable. She was not concerned with being liked. She was concerned with being irreplaceable.

Modern women in leadership face different battles but must adopt the same mindset. They cannot afford to lead as men do because they are not judged as men are. A male leader can be cold, ruthless, even cruel, and he will be respected for his strength. A female leader who exhibits the same traits will be seen as unnatural, as unlikable, as a problem that must be solved. But if she tries too hard to be warm, accommodating, or soft, she will be dismissed as weak, as lacking the decisiveness necessary to command. The world will never give her a perfect formula for how to lead, because the world does not want her to succeed. She must stop looking for one.

The balance between fear and respect is crucial. A woman who is feared but not respected will be seen as a tyrant, a temporary obstacle that must be removed. A woman who is respected but not feared will be taken advantage of, treated as someone whose authority is negotiable. The key is to create an environment where questioning her leadership is not an option. This does not mean ruling through terror or intimidation, it means ensuring that those around her understand that her authority is real, that there are consequences for disloyalty, that she is not someone who can be easily overthrown.

This is not about cruelty. It is about survival. A woman in power cannot afford to be naïve. She cannot afford to assume that her success will be enough to protect her. It will not. The higher she climbs, the more enemies she will have. The more she accomplishes, the more people will want to see her fail. She must be ready for this. She must anticipate betrayal before it happens. She must prepare for the inevitable backlash. She must never assume that she has won simply because she has reached the top. Holding power is harder than obtaining it, and she must be relentless in protecting what she has built.

Isolation is one of the greatest challenges women in power face. The higher they rise, the fewer true allies they have. Men in leadership naturally form networks, alliances, old boys' clubs that protect them even when they fail. Women are often alone. They are held to higher standards, expected to navigate the treacherous world of power without the same support systems that men enjoy. This is why they must be strategic in their alliances, careful in whom they trust, relentless in ensuring that they are never in a position where they can be easily removed.

Criticism will be constant. Every decision she makes will be questioned. Every move she makes will be analyzed. She will be expected to justify herself in ways men never have to. She must not let this distract her. She must not waste time trying to win over those who will never accept her authority. She must focus on consolidating power, on making herself indispensable, on ensuring that no matter how much resistance she faces, she remains in control.

The burden of expectation placed upon women in power is crushing, but it is also revealing. It exposes who is truly capable of leading. A woman who can withstand the relentless attacks, the endless scrutiny, the constant resistance, is a woman who has already proven herself stronger than the men who walk into leadership without a second thought. The world does not fear weak women. It does not sabotage them, does not attempt to remove them, does not spend time discrediting them. The world fears the women who refuse to be controlled.

The woman prince does not concern herself with fairness. She does not waste time trying to be everything to everyone. She does not dilute her authority to make others comfortable. She does not explain herself to those who have already decided that she should not lead. She rules. She leads with the knowledge that she will never be fully accepted, that there will always be those who resent her, that there will always be forces working against her. And she does not let it stop her.

She does not rule for approval. She rules because she has decided that she will not be moved. And that is what makes her unstoppable.

Part 3

A woman who rules in a world that fears her must prepare for a reality that most men in power will never have to face. The moment she asserts her authority, she is no longer just a leader, she is a threat. The system will not adapt to her presence; it will fight it. She will be scrutinized, challenged, tested, and undermined at every turn, not because she is incompetent, but because she is not meant to exist in the position she has claimed. Every move she makes will be questioned more harshly, every decision will be judged against an impossible standard, and every failure, real or manufactured, will be used as proof that women are unfit to lead. This is not paranoia. This is historical fact.

No woman who has ever ruled has done so without facing relentless opposition. The very act of holding power while being female is considered an offense by those who have spent centuries ensuring that power remains in male hands. This is why the woman prince must never be surprised by betrayal, by sabotage, by the constant attempts to undermine her. It is not personal; it is structural. She must not waste time expecting fairness or equality. She must operate with the understanding that no matter how competent she is, she will be resented. No matter how effective her leadership, she will be doubted. And no matter how much she proves herself, the system will never fully accept her.

This is not a reason to retreat. It is a reason to rule harder. Women who have survived in positions of power have done so by refusing to let opposition weaken them. They did not spend their time explaining themselves or trying to win over those who would never accept them. They focused on maintaining control, on strengthening their alliances, on ensuring that they were irreplaceable. A woman in power must never give her enemies the satisfaction of seeing her rattled. She must never let criticism dictate her actions. She must never allow herself to be pulled into the game of proving that she deserves to be there. The moment she does, she has already lost.

Elizabeth I did not waste time justifying her rule to those who believed a woman could never be a true sovereign. She acted as though her authority was unquestionable, and in doing so, she made it so. She understood that to lead as a woman, she could not appear vulnerable, uncertain, or hesitant. She created an image of herself as England's untouchable ruler, the Virgin Queen who belonged to no man and owed her power to no one but herself. She did not ask for loyalty, she commanded it. And when she faced betrayal, she did not hesitate to eliminate those who threatened her. She ruled with the knowledge that her enemies were always waiting for her to falter, and she ensured that she never gave them the opportunity.

Catherine the Great came to power through force, not inheritance. She did not pretend that she had been granted legitimacy, she took it. And once she was in control, she never apologized for how she had claimed her throne. She made herself indispensable to Russia, expanding its empire, reforming its institutions, and proving through sheer effectiveness that removing her would be more dangerous than keeping her. She did not concern herself with whether people liked her. She ensured that they needed her. She did not expect fairness. She expected resistance, and she prepared for it.

Cleopatra knew that as a woman, she would never be granted the same respect as a male ruler. She did not waste time trying to change that perception, she used it to her advantage. She understood that power was not just about ruling, but about ensuring that others saw her as the only one capable of ruling. She formed alliances strategically, aligning herself with powerful men not out of submission, but out of necessity. She presented herself not just as a queen, but as a goddess, an untouchable force whose rule was beyond question. She knew that the moment she allowed herself to be seen as just another leader, she would be vulnerable. So she made sure she was seen as something more.

The modern woman in power faces different obstacles, but the tactics remain the same. She will never be judged as a man is judged. She will never be given the benefit of the doubt. She will be tested, scrutinized, and attacked in ways that no male leader will ever experience. And she must be ready for it. She must not hesitate. She must not second-guess herself. She must rule with the knowledge that every move she makes will be met with resistance, and she must act accordingly.

She must also understand the burden of isolation. The higher she rises, the fewer true allies she will have. She must be careful in choosing whom to trust, because betrayal is not a possibility, it is a certainty. Men in power have built-in networks, alliances that protect them even when they fail. Women have no such safety net. They must construct their own, ensuring that their power is not dependent on the approval of those who seek to undermine them.

A woman in leadership must also recognize that her failures will be magnified while her successes will be minimized. If she makes a mistake, it will be treated as proof that women are not fit to lead. If she succeeds, it will be attributed to luck, to the help of male advisors, to circumstances outside of her control. She will never receive full credit for what she accomplishes, and she must stop expecting it. Her focus must not be on receiving recognition, but on maintaining control.

Criticism will come from every direction. It will not always be about her leadership, it will be about her voice, her appearance, her demeanor, her personal life. She will be criticized for being too harsh, too emotional, too cold, too ambitious. The contradictions are designed to be impossible, to keep her constantly adjusting, constantly doubting, constantly fighting battles that men never have to fight. The only way to win is to stop playing the game. She must not let herself be consumed by proving that she is good enough. She must act as though she already is.

The world will never fully accept a woman in power. This is the reality. But a woman who understands this, who does not waste time seeking approval, who does not dilute her authority to make others comfortable, who does not explain herself to those who will never respect her, is a woman who cannot be stopped. She does not rule for acceptance. She rules because she has claimed power, and she has no intention of giving it up. The woman prince does not rule for applause. She does not rule to be loved. She rules because she has decided that she will not be moved. And that is what makes her unstoppable.

~6
Reputation

The Double-Edged Sword

Reputation is both an asset and a weapon. It can be cultivated, manipulated, and used to build unshakable authority, or it can be turned against a woman in power to discredit, humiliate, and ultimately remove her. No woman in history has ever ruled without being subject to ruthless attacks on her character. A man's failures in leadership are often attributed to poor decision-making, incompetence, or political miscalculation. A woman's failures, however, are always framed as personal deficiencies, her character is flawed, her emotions unstable, her ambition unnatural. She is not just a bad leader; she is a bad woman. This is why reputation must never be treated as an afterthought. It is not just about how a woman in power is seen, it is about whether she is allowed to lead at all.

Women have long been told that they must "play nice" to succeed, that they must earn respect by proving themselves beyond question, that they will be granted power if only they are patient and cooperative. This is a lie. Women who wait to be granted legitimacy are left waiting forever. Women who seek power must take it, and they must control the way they are perceived in the process. The myth of playing nice is one of the most effective tools of control ever used against women. It convinces them to soften themselves, to hesitate, to worry more about being liked than about being effective. It keeps them preoccupied with approval rather than strategy. But history does not reward women who wait politely for recognition. It rewards women who define their own image before others do it for them.

A woman's reputation is never neutral. It is always either being strengthened or being used against her. And in today's world, the control of reputation has become more difficult than ever. The rise of mass media, social media, and digital communication has made it easier than ever for reputations to be weaponized. A single headline, a single misinterpreted phrase, a single carefully placed accusation can undo years of work. Smear campaigns are not accidents; they are deliberate tools of destruction used against women who refuse to conform, who threaten the established order, who refuse to be silent. A woman in power must never be naïve about this. She must expect it, prepare for it, and control her own narrative so completely that any attack on her reputation is ineffective before it even begins.

Reputation has been one of the most effective weapons used against powerful women throughout history. Women in leadership are rarely attacked solely on the basis of their policies or decisions. Instead, the attacks are personal. Their morality is questioned. Their ambition is framed as dangerous. Their personal lives become public battlegrounds, with every mistake, real or imagined, amplified to ensure that they are seen not as leaders, but as unfit, unstable, unworthy of power. This tactic is as old as power itself.

Consider Cleopatra. She was one of the most brilliant political minds of her time, a ruler who ensured Egypt remained independent despite the growing dominance of Rome. She was a strategist, a diplomat, a leader who understood the necessity of alliances and perception. And yet, how does history remember her? Not as a ruler, but as a seductress. Not as a strategist, but as a woman who used her beauty to manipulate men. Her intelligence was erased, her leadership reduced to the idea that she relied on sexual appeal rather than political skill. This was not an accident. It was a deliberate rewriting of history, a way to diminish her legacy, to ensure that she was remembered not as a woman who ruled but as a woman who was defined by the men around her.

Catherine the Great faced similar attacks. She ruled Russia for over three decades, expanding its borders, modernizing its government, strengthening its economy. She was one of the most effective leaders in Russian history. But what do people remember? The salacious rumors, the lies about her personal life, the absurd accusations that had nothing to do with her leadership but everything to do with the need to discredit a woman who had seized power and refused to let it go. Her reputation was used against her in death just as it had been used against her in life, and yet, she remains one of the most enduring figures in history because she never let those attacks dictate her rule.

Elizabeth I understood that reputation was not just a matter of perception, it was a matter of survival. She ruled in a time when a woman on the throne was seen as an anomaly, a problem that needed to be corrected. She knew that if she married, she would be diminished, reduced to the status of a wife rather than a sovereign. So she cultivated an image of herself as the Virgin Queen, a ruler who belonged only to England, untouchable and beyond the reach of ordinary expectations. She did not let others define her, she created her own legend, ensuring that even those who despised her could not question her authority.

The lesson from these women is clear: reputation is not something to be left to chance. It must be controlled, shaped, and used as a weapon before it can be turned against you. A woman in power cannot afford to assume that her work will speak for itself. It will not. She must ensure that her narrative is so strong, her image so carefully constructed, that no attack on her reputation can destroy her. She must anticipate how she will be perceived, and she must use that perception to her advantage.

There are times when embracing the role of the villain is the only option. When a woman in power is determined to be unlikable no matter what she does, when she is framed as too ambitious, too aggressive, too ruthless, she has two choices. She can waste time trying to soften herself, trying to prove that she is good, that she is kind, that she is acceptable. Or she can lean into the fear, ensure that those who seek to undermine her understand that she will not be controlled. A woman who is feared is a woman who is difficult to remove. A woman who is necessary is a woman who cannot be ignored.

This does not mean being cruel for cruelty's sake. It means understanding that power does not come from being liked, it comes from being respected. A woman who is respected will always be more powerful than a woman who is merely liked. When her enemies call her difficult, she must recognize that what they truly mean is that she is not compliant. When they call her ruthless, they mean that she is not easily controlled. When they call her unlikable, they mean that she has refused to shrink herself to make them comfortable.

Modern women in leadership must take these lessons and apply them ruthlessly. They must understand that their reputation will always be a battleground, that their image will always be scrutinized, that they will always be judged more harshly than the men around them. They must not let this deter them. They must control the story before others can write it for them. They must understand that reputation is a double-edged sword, one that can be turned against them in an instant, but one that can also be wielded in their favor if they are strategic enough.

A woman in power must decide what she wants to be known for. She must choose her image with purpose. She must ensure that no matter what attacks come her way, her authority remains intact. She must recognize that there will be those who will always see her as a villain, and she must not let that weaken her. Because in the end, history remembers those who rule, not those who beg for acceptance. And the women who have ruled, truly ruled, have never wasted time trying to be liked. They have understood that their power was never meant to be granted. It was meant to be taken. And they have taken it without apology.

Part 2

Reputation is not just an accessory to power, it is the battlefield on which a woman's authority is won or lost. A man in power can afford to be careless with his image, knowing that his legitimacy is rarely questioned. A woman in power, however, is never given that luxury. Her reputation is not simply a reflection of her leadership; it is a tool that will be used against her at every opportunity. If she does not control her own narrative, others will do it for her, and they will not do it kindly.

For women in leadership, reputation is not a passive force. It is an active weapon, one that can be sharpened and wielded or dulled and turned against its owner. Women who assume that their work alone will define them, that their actions will be enough to secure their place, that fairness will prevail if they simply remain competent and disciplined, are setting themselves up for failure. A woman who does not define herself will be defined by those who seek to undermine her. She will be framed as too ambitious, too difficult, too cold, too emotional, always too much or not enough. The contradictions are endless, and they are deliberate. They exist to keep her in a state of constant defensiveness, to ensure that she is always reacting rather than leading.

The most powerful women in history have understood this, and they have refused to allow their reputations to be shaped by others. They have taken control of their own narratives, ensuring that they are seen in exactly the way they need to be in order to hold onto power. This is not about honesty; it is about survival. A ruler who believes that reputation is merely a reflection of truth is a ruler who will not last long. Reputation is not about what is real. It is about what is perceived, and perception, in the world of power, is everything.

Elizabeth I mastered the art of self-mythologizing. She did not allow herself to be seen as merely a queen, she made herself into a legend. By crafting the image of the Virgin Queen, she ensured that she was untouchable, beyond the reach of the men who might otherwise have tried to dominate or control her. She did not allow herself to be reduced to the status of a wife or a pawn in diplomatic marriages. She presented herself as a ruler who belonged only to England, an entity whose authority could not be questioned. She controlled her own image so effectively that even centuries after her death, she remains one of the most iconic figures in history. She did not waste time trying to convince people of her worth, she ensured that they had no choice but to accept it.

Catherine the Great followed a similar strategy, though she faced even greater opposition. Unlike Elizabeth, who inherited her throne, Catherine took hers. She knew that she would never be fully accepted as a legitimate ruler, that her foreign birth and the nature of her rise to power would always be used against her. So she made herself indispensable. She focused not on trying to change the opinions of her critics but on ensuring that even those who despised her had no choice but to acknowledge her effectiveness. She modernized Russia, expanded its territory, strengthened its economy. She did not concern herself with whether people liked her. She concerned herself with ensuring that they needed her.

Cleopatra understood that in a world where female rulers were the exception, not the rule, perception was as powerful as any army. She did not merely rule Egypt, she embodied it. She presented herself not as a mere queen but as a living goddess, ensuring that her authority was seen as divine rather than political. She aligned herself strategically with powerful Roman figures not out of submission, but out of necessity, ensuring that Egypt remained a key player in the most dominant empire of its time. And yet, her reputation was rewritten by history, reduced to the image of a seductress rather than a ruler. This was not an accident. It was a deliberate effort to diminish her, to erase her political intelligence and strategic brilliance, to ensure that she was remembered not as a leader but as a woman who had used men to gain power.

This is one of the greatest dangers of reputation for women in power. Their successes are rewritten, reframed, stolen. Their leadership is attributed to luck, to the influence of men, to factors beyond their control. They are treated not as rulers in their own right, but as anomalies, exceptions to the rule that power belongs to men. And their reputations are always at risk of being turned against them, of being used to justify their downfall. A woman in power must never assume that her legacy will be written fairly. She must write it herself.

Modern women in leadership face the same battles, though the tools have changed. The media has become the ultimate weapon in the war over reputation. A single article, a single scandal, a single moment of perceived weakness can undo years of careful image-building. Smear campaigns are not accidents, they are deliberate efforts to discredit women who have become too powerful, too influential, too difficult to control. And social media has only amplified the ability of these attacks to spread, turning reputation into something that can be destroyed in an instant.

This is why a woman in power must control her own narrative before anyone else can. She must anticipate how she will be perceived and shape that perception to her advantage. She must be ruthless in ensuring that her authority is not undermined by false narratives, by misinterpretations, by the endless attempts to paint her as something she is not. And she must understand that there will be times when she must embrace being the villain.

A woman in power will always be framed as too much. She will always be seen as threatening to someone. If she tries to please everyone, she will fail. She must choose how she will be feared. She must decide what kind of reputation will serve her best. She must recognize that in a world that does not want her to lead, she will be painted as ruthless, ambitious, and difficult no matter what she does. And rather than fighting against these labels, she must decide how to use them.

There is strength in being seen as formidable. There is power in being feared. A woman who is feared is a woman who is difficult to remove. A woman who is respected is a woman whose authority cannot be easily questioned. A woman who is both is unstoppable.

This does not mean that a woman in power must be cruel or unfeeling. It means that she must be strategic. She must understand when to be harsh and when to be merciful, when to push forward and when to retreat. She must ensure that her reputation is never left to chance, that it is never dictated by those who seek to destroy her. She must craft her own legend, as Elizabeth did. She must make herself necessary, as Catherine did. She must make herself larger than life, as Cleopatra did.

And she must be prepared for the inevitable attacks. She must expect that no matter how much she accomplishes, there will be those who try to rewrite her history, who try to frame her success as something that was given to her rather than something she took. She must anticipate the backlash, the rumors, the attempts to discredit her. And she must make it so that none of it matters.

A woman who controls her reputation controls her power. She must not waste time proving that she deserves to be in the room. She must act as though it is already hers. She must understand that she will always be framed as something, so she must choose what that will be. She must decide whether she will be seen as a figure to be dismissed or a force to be reckoned with.

History does not remember women who waited to be accepted. It remembers women who ruled despite never being fully embraced. It remembers those who refused to apologize for their ambition, who did not waste time softening themselves to be more palatable, who did not allow their reputations to be written by their enemies.

The woman in power does not wait for approval. She does not waste time trying to win over those who will never respect her. She shapes her own legend. She ensures that when history remembers her, it is on her own terms. And she rules, not because she is allowed to, but because she has made it impossible for anyone to stop her.

Part 3

Reputation is not simply a reflection of leadership, it is the battlefield where power is won or lost. A man in power can afford to be careless with his image, knowing that his legitimacy is rarely questioned. A woman in power is never given that luxury. Her reputation is both a tool and a liability, a shield that can protect her and a sword that can be turned against her at any moment. There is no neutrality in how she is perceived. Every action, every decision will be scrutinized not just for its impact but for what it suggests about her as a woman in power. A man in leadership can fail, recover, fail again, and still be granted legitimacy. A woman does not have that privilege. One misstep, one moment of perceived weakness, one scandal, real or manufactured, can erase everything she has built.

A woman in power must never leave her reputation to chance. She cannot assume that her work alone will define her. She must actively shape how she is seen, control the narrative before it is written for her, and anticipate the inevitable attacks on her character. If she does not, she will find herself fighting a battle she cannot win, where the rules are rigged and the outcome decided by those who would rather see her fail.

The world has always been eager to tear down powerful women. It does not matter how competent they are, how much they accomplish, or how necessary their leadership might be. The moment they refuse to conform to expectations, they become targets. And the most effective weapon used against them is reputation. History is filled with stories of women who were vilified, slandered, and reduced to caricatures designed to erase their actual impact. The more successful a woman becomes, the more determined the world is to rewrite her legacy.

Cleopatra ruled Egypt with intelligence and strategy, ensuring that her kingdom remained dominant in an era of Roman expansion. Yet she is remembered not as a brilliant leader but as a seductress who used her beauty to manipulate men. The message is clear: a woman who wields power effectively must have done so through deception rather than skill. Catherine the Great transformed Russia into one of the most powerful nations in the world, but her reputation was tarnished by scandal and slander. Her personal life was exaggerated to overshadow her accomplishments. Elizabeth I controlled her image so tightly that no attack on her character could truly take hold. She made herself into a myth, the Virgin Queen, untouchable and beyond expectation.

The lesson from these women is clear: reputation is not something a woman in power must merely manage. It is something she must weaponize. She cannot afford to be passive about how she is seen. She must anticipate how her enemies will seek to discredit her and ensure that their attacks are ineffective before they even begin. She must create a persona so powerful that any attempt to undermine her only reinforces her authority.

A modern woman in leadership faces the same fundamental challenge. The media, social media, and digital platforms have made it easier than ever to attack a woman's reputation, to spread lies, to create narratives that are difficult to undo. A single misstep can be amplified, a single moment of human error turned into a scandal that defines her career. The expectations placed upon her remain impossible. If she is too strong, she is called aggressive. If she is too measured, she is weak. If she is too ambitious, she is ruthless. If she is not ambitious enough, she is ineffective. The contradictions are intentional, they keep her trapped in self-doubt, reacting instead of leading.

She must stop playing by the rules that were not written for her success. She must accept that she will never be universally liked. She must recognize that she will always be framed as "too much" in one way or another. A woman in power will never please everyone, so she must stop trying. Instead, she must ensure that her authority is undeniable, that her reputation serves her interests, and that any attack on her only strengthens her position.

There are times when embracing the role of the villain is necessary. The most dangerous thing a woman in power can do is try to make herself palatable to those who will never accept her. A woman seen as ruthless is often simply a woman who refuses to be controlled. A woman called difficult is often just one who does not conform to expectations designed to keep her small. A woman feared is not easily removed.

This does not mean cruelty is strength. It means a woman in power must be strategic in how she cultivates her image. She must understand when to be harsh and when to be measured, when to allow rumors to spread and when to crush them. She must know that reputation is not about truth, it is about perception. And perception, in the world of power, is often the only thing that matters.

She must decide what she will be known for. Will she be seen as a ruler who cannot be challenged? Will she be a force of stability, a leader who is indispensable? Will she make herself into a myth, something larger than life, untouchable by ordinary attacks? These choices must be made deliberately. A woman in power who does not shape her reputation will find herself at the mercy of those eager to define her in the worst possible terms.

History does not remember women who played it safe. It does not remember those who tried to be liked, who sought approval, who bent themselves into impossible shapes to satisfy expectations that were designed to be unattainable. It remembers the women who ruled despite the resistance, who did not apologize for their ambition, who refused to let their reputations be dictated by those who wanted to see them fall.

A woman in power does not wait for her reputation to be given to her. She takes it. She defines it. She ensures that it serves her rather than controls her. She does not shrink to make others comfortable. She does not waste time explaining herself to those who have already decided that she should not lead. She does not rule for approval. She rules because she has made it impossible for anyone to stop her.

~7
The Illusion of Fairness

Why Women Must Play the Long Game

Power has never been about fairness. It has never been granted to those who deserve it, nor has it been distributed according to merit or moral standing. It is taken, consolidated, and defended by those who understand its rules, and those rules have never favored women. A woman who seeks power must accept this reality: fairness is an illusion, a story told to keep her hopeful, to keep her waiting, to keep her compliant. If she expects justice, if she believes that patience alone will grant her authority, she will spend her life watching lesser men rise while she remains locked outside of the decision-making rooms.

Women have been told for centuries that change will come if they simply prove themselves enough, if they endure with grace, if they demonstrate patience and resilience. They are told to wait, to be reasonable, to trust that the system will correct itself in time. This is a lie. The system does not correct itself, it is corrected by those who refuse to accept its terms. Every right, every freedom, every inch of progress that women have gained throughout history has come not from waiting, not from playing by the rules, but from challenging them, from disrupting the existing order, from refusing to accept a world in which their power is secondary. And yet, the illusion of fairness persists, used as a weapon to keep women in line, to prevent them from demanding what they are owed, to convince them that the game is not rigged against them when it always has been.

A woman who seeks real power must abandon the belief that if she follows the rules, she will be rewarded. The rules were never made for her success. If they were, she would not be forced to work twice as hard for half the recognition. She would not be expected to prove herself endlessly while men are allowed to fail upward. She would not be judged more harshly, scrutinized more unfairly, and dismissed more easily. The system is designed to exhaust her, to make her believe that the problem is her rather than the structures that were built to exclude her. This is why she must stop waiting. She must stop believing that fairness will one day prevail. She must embrace the long game, the understanding that change is not about waiting for the right moment, it is about creating the moment, about forcing it into existence when the world refuses to grant it.

Systemic change does not happen in election cycles. It does not happen in a single term, a single movement, a single moment of collective action. It happens over decades, sometimes centuries, built by those who understand that progress is not immediate, that real transformation requires persistence beyond what most are willing to endure. The women who have reshaped history did not always see the results of their work in their lifetimes. Some died before their movements succeeded, before their ideas became mainstream, before their victories were recognized. But they fought anyway, knowing that they were laying the foundation for something greater than themselves.

Consider the suffragists, the women who spent their lives demanding the right to vote, who were mocked, imprisoned, beaten, and ignored for decades before their efforts bore fruit. Many of them did not live to see their victories realized. They fought not because they knew they would win in their time, but because they understood that someone had to start the battle, that progress required sacrifice, that the world does not change unless people force it to.

The same is true of the women who fought for civil rights, for labor protections, for reproductive freedom. They did not operate under the illusion that fairness would eventually grant them justice. They understood that power concedes nothing willingly. They knew that their victories would not come overnight, that setbacks were inevitable, that backlash was guaranteed. But they persisted, because they knew that even if they did not see the results of their struggle, others would.

This is the mindset a woman must embrace if she is serious about power. She must think beyond her own lifetime, beyond her own immediate goals, beyond the next election, the next promotion, the next personal success. She must see the long game, the work that stretches beyond what she alone can accomplish, the battles that will not be won in a single generation but that must be fought anyway. This is not about individual victories. It is about building something that cannot be undone, something that survives beyond the temporary wins and losses, something that reshapes the very structures that have kept women out for so long.

A woman who plays the long game does not waste time seeking validation from a system that was never built for her. She does not expect recognition for every step forward, nor does she allow setbacks to deter her. She understands that resistance is part of the process, that those in power will always fight against those who seek to change the order of things. She does not mistake opposition for failure. She sees it as proof that she is on the right path.

This is the lesson history teaches again and again: the women who persist are the women who win. They may not see their victories immediately. They may be ridiculed, dismissed, cast aside. But their work does not disappear. Their ideas do not vanish. They plant the seeds of change, and those seeds grow, even when they cannot see them take root.

Power is not about fairness. It is about persistence. It is about understanding that the fight does not end simply because it is exhausting, because it is difficult, because it seems endless. The world will not hand women what they deserve. It never has.

But those who refuse to accept the illusion of fairness, those who play the long game, those who keep pushing forward even when progress seems impossible, these are the women who shape history.

A woman who seeks real power must think beyond herself. She must understand that her fight is bigger than her own ambitions. She must build something that lasts, something that cannot be easily dismantled, something that outlives those who try to stop it. She must see every battle, every challenge, every setback not as a reason to retreat but as confirmation that she is part of something larger than the present moment.

History does not remember those who waited for fairness. It remembers those who refused to accept a world that was not built for them and who reshaped it anyway. And a woman who understands this, who does not seek permission, who does not wait to be granted what she deserves, who does not give up simply because the fight is long, is a woman who will win. If not in her lifetime, then in the world she leaves behind.

Part 2

Power is a long game, and women who seek it must embrace the reality that real change is not instant, not easy, and certainly not fair. The world has never willingly handed power to those it seeks to exclude, and it will not start now. Those who expect fairness will be waiting forever. Those who believe the system will correct itself in their favor will find themselves watching from the sidelines as others, less qualified, less capable, but more ruthless, take the positions they were told they could earn if only they were patient. The lesson history teaches, over and over, is that patience is not enough. Power is not about waiting. It is about persistence, endurance, and the ability to outlast those who seek to keep it from you.

Systemic change has never been won in a single moment. The great revolutions, the transformative movements, the seismic shifts that have restructured societies did not happen because one person or one generation willed them into being. They happened because of the relentless, often thankless work of those who understood that progress is not linear, that setbacks are inevitable, and that victories are sometimes claimed by those who did not begin the fight. This is the reality that women in power must accept. They are part of a long chain of those who came before them and those who will come after them. Their work is not just about the present, it is about shaping the world in ways that may not be fully realized until long after they are gone.

Women have been told for centuries that if they are good enough, smart enough, patient enough, they will be allowed in. This is the greatest lie ever told to keep them complacent. The system is not designed to reward women for their excellence. It is designed to exclude them while offering just enough incremental progress to keep them hopeful. This is why women must reject the illusion of fairness and instead embrace the long game, the knowledge that real power is not won by those who expect it to be handed to them, but by those who refuse to stop fighting for it.

The suffragists did not win the right to vote in a single campaign. It took generations of women enduring arrests, force-feedings, beatings, and relentless political obstruction before suffrage was achieved. Many of the early fighters for women's rights did not live to see their victories realized. They fought not because they expected immediate results, but because they knew that change required persistence beyond their own lifetimes.

The women who fought for labor rights, for reproductive freedom, for civil rights did not achieve their goals overnight. They were resisted, ridiculed, and often attacked by the very institutions that claimed to value justice and democracy. Their victories were hard-won, and even after they succeeded, their progress was continually threatened, undermined, and reversed by those who refused to accept that women could hold real power. But still, they fought. And because they fought, because they refused to accept that they had to wait for permission, the world changed.

A woman who seeks power today must recognize that she is part of this history. She must understand that the barriers she faces are not new, that the resistance she encounters is not unique to her, and that the struggle she endures is one that countless women before her have faced. This is not a reason to despair. It is a reason to push forward, to keep going even when the path ahead seems impossible.

The world will tell her that she is being too ambitious, too impatient, too demanding. It will tell her that change takes time, that she must wait her turn, that she must be careful not to push too hard or she will alienate those who might support her. These are lies. Change does not happen because people wait their turn. It happens because people refuse to wait any longer.

Women who lead movements, who rise to power, who break barriers must stop believing that they will be granted a seat at the table if only they behave correctly. The table was not built for them. The rules were not written for their success. They must take power, seize it in ways that make it impossible to ignore them, and build structures that will outlast the forces working against them.

History is filled with women who understood this, who refused to play by the rules of fairness and instead rewrote the rules themselves. Harriet Tubman did not ask permission to free enslaved people, she did it, knowing that the law was against her, knowing that the system was not designed to allow her to succeed. Sojourner Truth did not wait for someone to give her a platform, she took it, speaking to crowds who doubted her right to even have a voice.

More recently, women in politics, business, and activism have made it clear that playing by the existing rules will never be enough. Women like Angela Merkel, who held power in Germany longer than most of her male predecessors, did not succeed by waiting for permission. She consolidated authority, built strategic alliances, and ensured that she was indispensable. Ruth Bader Ginsburg did not rise to the Supreme Court by assuming the system would naturally allow women into its highest ranks, she fought for every inch of ground she gained, knowing that if she did not, the door would remain closed to those who followed.

A woman who plays the long game does not get distracted by small defeats. She does not allow short-term setbacks to break her resolve. She knows that power is not about immediate victories, it is about ensuring that the fight continues, that the movement survives, that the system is reshaped whether or not she sees the final result.

The world will try to exhaust her. It will throw obstacles in her path, force her to justify her existence, make her prove her worth again and again while letting mediocre men ascend without question. This is not evidence that she is failing, it is evidence that she is doing something right. The more resistance she faces, the more she must recognize that she has become a threat to those who wish to maintain the existing order.

She must embrace this role. She must understand that power does not accommodate those who seek to change it. It resists them, undermines them, tries to outlast them. And so, she must refuse to be outlasted. She must refuse to accept that her time will come later. She must recognize that every moment she waits is a moment wasted, a moment in which the forces working against her continue to strengthen their hold.

Playing the long game does not mean accepting delays. It does not mean waiting. It means understanding that every action, every move, every battle is part of a larger war. It means knowing that change is won through persistence, through endurance, through an unshakable belief that no matter how hard the fight, it must be fought anyway.

A woman who seeks power must train herself to see beyond the present. She must understand that every step forward is part of a larger path, that even when she does not see immediate success, she is laying the groundwork for those who will come after her. She must understand that movements do not belong to individuals, that victories are often collective, that true power is not about one person winning, but about reshaping the world in ways that cannot be undone.

The lesson of history is that those who persist are those who win. The suffragists were told to wait, and they refused. The civil rights activists were told that the time was not right, and they refused. Women who have fought for justice, for rights, for representation have always been told to be patient, to wait for the right moment. And they have always won by refusing to wait.

A woman who understands this, who does not waste time hoping for fairness, who does not wait for an invitation to power, who does not accept that the rules must be followed as they are written, is a woman who will win. Maybe not today, maybe not in her own lifetime, but in the legacy she leaves, in the change she sets in motion, in the world that will be different because she refused to accept that she had to wait.

The long game is not for the weak. It is not for those who need immediate gratification. It is for those who understand that the greatest victories are not always seen by those who begin the fight, but by those who inherit their work. A woman who seeks power must embrace this truth. She must keep moving forward, keep pushing, keep fighting, knowing that fairness will never be handed to her, that justice will never come without struggle, that the world will only change if she forces it to.

And that is why she must never stop.

Part 3

Power is never about fairness, and those who wait for fairness will find themselves waiting forever. A woman who seeks to lead, who dares to claim authority in a world designed to keep her out, must accept that the system will never willingly make room for her. She must reject the illusion that if she plays by the rules, if she waits her turn, if she proves herself beyond question, she will be rewarded. She will not. The game is not fair, and it never has been. And yet, the world tells women to be patient, to wait, to believe that if they endure long enough, their time will come. This is a lie. History has never rewarded patience. It has rewarded those who refuse to accept the terms imposed upon them.

A woman who plays the long game understands this. She does not waste time waiting for validation from a system that was built to exclude her. She does not wait for permission. She takes power because she understands that the only way to create real change is to force it. The great movements led by women throughout history were not successful because the world suddenly decided to be fair. They were successful because the women leading them refused to back down, refused to compromise, refused to accept that their demands were unreasonable. They understood that power is not about deserving it, it is about taking it and holding onto it long enough to transform the world in their image.

The long game is brutal. It requires patience, but not the kind of patience that leads to complacency. It is not about waiting, it is about enduring. It is about pushing forward even when progress is slow, even when resistance is overwhelming, even when the goal seems impossibly far away. It is about recognizing that real power is not won in a single moment, not in a single election, not in a single protest, but in the relentless, grinding work of persistence. The world does not change easily. It does not give up power without a fight. But history belongs to those who are willing to keep fighting, no matter how long it takes.

Systemic change has never been a matter of fairness. It has been a matter of force, of strategy, of refusing to accept defeat even when victory seems impossible. The suffragists did not win the right to vote because the world suddenly decided that women were equal. They won because they refused to be ignored. They won because they disrupted the system, because they made it impossible to continue excluding them without consequence. They were mocked, beaten, imprisoned, force-fed, dismissed as radicals. But they did not stop. They understood that the fight was bigger than them, that even if they did not live to see their victories, they were laying the groundwork for those who would come after them.

The women who fought for civil rights, for labor protections, for reproductive freedom, for representation in government and business, did not win their battles because they were patient. They won because they were relentless. They won because they refused to be silenced, because they kept pushing forward even when the opposition was overwhelming. And even now, their victories are not permanent. Every gain that women have made has been met with backlash, with attempts to roll back progress, to return to a world where power remains concentrated in the hands of men. This is why the fight can never stop. A woman who seeks power must understand that her work is never done.

She must also understand that playing the long game does not mean sacrificing ambition in the present. It does not mean settling for small, incremental progress. It means thinking beyond herself, beyond her own lifetime, beyond the immediate victories and losses. It means building something that lasts, something that cannot be undone, something that reshapes the world in ways that extend far beyond her own influence. This is how real power is won, not through individual success, but through the creation of systems that ensure that once power is taken, it cannot be taken back.

Women who lead today must take this lesson seriously. They must recognize that every battle they fight is part of a larger war, that even when they lose in the moment, they are contributing to a movement that will outlast them. They must stop thinking in terms of short-term wins and start thinking in terms of generational transformation. They must be willing to push for change that they may never see fully realized. They must be willing to fight for a world they will not live in, knowing that the work they do now will make that world possible.

This is what the long game requires: the ability to endure, to keep moving forward, to accept that power is not won overnight but through years, decades, even centuries of struggle. It requires resilience, the understanding that setbacks are inevitable, that progress is never linear, that every step forward will be met with resistance. It requires the ability to see beyond the immediate, to recognize that the work of one woman, one movement, one generation is part of something much larger, something that stretches across time, something that cannot be undone if enough people commit to seeing it through.

A woman who plays the long game must also be strategic. She must understand that power is not just about taking action, it is about knowing when to act, when to push forward, when to hold back, when to strike. She must anticipate the opposition, understand their tactics, prepare for the inevitable backlash. She must know that every move she makes will be scrutinized more harshly than any man in her position, and she must refuse to let that scrutiny dictate her choices. She must not waste time defending herself against accusations that she is too ambitious, too aggressive, too much. She must embrace the reality that she will always be too much for those who wish to see her fail, and she must decide to be too much anyway.

This is how history is made. Not by those who wait, but by those who persist. By those who understand that power is not about fairness, but about the ability to outlast those who would rather see them disappear. The long game is brutal, but it is the only game worth playing. And a woman who plays it well will not just change her own life, she will change the world.

~8
Fear vs. Love

Leading Without Apology

Power is not about being liked. This is the first and most crucial lesson a woman who seeks to lead must learn. A man in power can afford to be disliked. He can be ruthless, cunning, even outright cruel, and he will still be respected, still be treated as legitimate, still be given the benefit of the doubt. A woman in power, however, is expected to be likable. She is expected to be warm, accommodating, endlessly patient. She is expected to lead not through command, but through persuasion. She is expected to make people comfortable, to ensure that they do not feel threatened by her authority, to prove that she is not a danger to the existing order. And if she refuses to play this role, if she dares to lead without apology, she is branded as difficult, as unlikable, as a tyrant.

Machiavelli understood that power is not about fairness, that it is better to be feared than loved, that those who rely on affection rather than control will find themselves abandoned when they need support the most. And in many ways, he was right. A ruler who is only loved has no real authority. A ruler who is only feared is vulnerable to rebellion. The most successful leaders throughout history have been those who understood how to balance these forces, how to command both admiration and intimidation, how to ensure that they were respected above all else.

But for women, the balance is far more complicated. A male leader who is feared is seen as strong. A female leader who is feared is seen as monstrous. A male leader who is admired is seen as inspiring. A female leader who is admired is seen as weak. The contradiction is impossible, and it is deliberate. It is designed to ensure that no woman in power can ever truly be safe, that she is always walking a tightrope, always adjusting, always trying to be just likable enough to avoid rebellion but just firm enough to maintain control.

This is the trap of likeability. It is a cage designed to keep women in powerless positions, to ensure that they are never truly able to lead without compromise. It is what keeps women endlessly trying to prove themselves, endlessly adjusting their tone, endlessly softening their authority so that they do not appear threatening. But power is threatening. Power is disruptive. Power is about making people uncomfortable, about forcing change, about ensuring that those who resist transformation do not have the ability to stop it. A woman who is afraid of making people uncomfortable will never hold real power. She will always be negotiating for scraps rather than commanding what is rightfully hers.

The most successful women in history have understood this. They have not wasted time trying to be liked. They have not softened their leadership to make others comfortable. They have ruled with the knowledge that they would never be fully accepted, never be fully embraced, never be given the same deference that was automatically granted to men. And they have ruled anyway.

Elizabeth I did not waste time trying to make herself likable. She knew that if she attempted to lead as a woman, she would be dismissed. So she led as something beyond a woman, she led as a legend, as a ruler whose authority was unquestionable. She did not ask for permission to govern; she governed. She made it clear that those who opposed her would be crushed, that her authority was absolute, that she would not be swayed by the demands of those who wished to see her fail. She was not interested in being liked. She was interested in being obeyed.

Catherine the Great ruled with the same mindset. She did not concern herself with whether the Russian nobility approved of her. She concerned herself with ensuring that they needed her. She modernized the empire, expanded its reach, secured its economy. She ensured that even those who despised her had no choice but to acknowledge that she was the best ruler Russia had ever seen. She did not beg for respect. She took it. And when she faced opposition, she did not hesitate to eliminate those who sought to challenge her authority.

Cleopatra understood that in order to survive, she could not afford to be seen as just another queen. She had to be something more. She had to be a force, a living goddess, a ruler whose power was unquestionable. She did not seek to be loved. She sought to be irreplaceable. And in doing so, she ensured that Egypt remained independent for as long as possible, that her rule was secure even in the face of overwhelming opposition.

But these women also understood that ruling through fear alone was not enough. They understood that terror without loyalty was unsustainable, that cruelty without admiration would lead to rebellion. They did not rely on fear alone. They ensured that their people saw them as indispensable, as the only ones capable of leading, as figures who were not just powerful but necessary. They were feared, but they were also respected. And this is the balance that modern women in leadership must master.

A woman who is only feared will be removed at the first opportunity. A woman who is only loved will never be taken seriously. The key is to be respected, to ensure that those who follow you do so not just out of obligation but because they understand that you are the only one capable of leading. This means setting clear expectations, enforcing consequences for disloyalty, ensuring that your authority is never questioned. But it also means inspiring confidence, ensuring that those who follow you believe in your vision, that they see you as someone who is not just ruling for the sake of ruling, but for the sake of something greater.

Modern women in power must stop wasting time trying to be liked. They must stop apologizing for their authority. They must stop softening their leadership in an attempt to avoid criticism. The criticism will come no matter what they do. They will always be seen as too much or not enough. They will always be scrutinized more harshly, judged more unfairly, held to impossible standards. The only way to win is to stop playing the game, to stop seeking approval from those who will never give it, to lead as though their authority is already unquestionable.

This does not mean being cruel. It does not mean ruling through terror. It means ensuring that respect is never negotiable, that those who seek to undermine them are not given the opportunity, that their leadership is seen as necessary rather than optional. It means commanding authority without hesitation, without apology, without the endless justifications that women are expected to provide.

Women who lead today must study those who came before them, must understand how they maintained control, how they balanced fear and admiration, how they ensured that their authority was not just accepted but unchallengeable. They must recognize that they will never be given the same freedom to fail that men are given. They must be relentless, unshakable, immune to the demands that they be softer, more likable, more accommodating.

Because in the end, power is not about being liked. It is about being respected. And a woman who is respected is a woman who cannot be removed. She is a woman who is not negotiating for her authority but commanding it. She is a woman who is not waiting for validation but enforcing her vision. She is a woman who understands that leadership is not about fairness, that it is not about pleasing people, that it is not about being acceptable. It is about ruling.

And a woman who understands this, who embraces it without hesitation, is a woman who will win. Not because she was allowed to, not because she was invited in, but because she refused to be denied.

Part 2

Power has never been about being liked. It has never been about fairness, about playing by the rules, about waiting for approval. Those who lead do not do so because they are loved. They lead because they are respected, because they command authority, because they make it clear that they will not be ignored. A woman who seeks power must understand this. She must stop trying to be likable. She must stop apologizing for her ambition. She must recognize that the world does not reward those who wait politely to be acknowledged. It rewards those who take what they need and refuse to let go.

Machiavelli was right: it is better to be feared than loved. But for women, the equation is more complicated. A man in power who is feared is seen as strong. A woman in power who is feared is seen as dangerous. A man who is respected for his ruthlessness is called decisive, authoritative, a leader. A woman who does the same is called cruel, heartless, a tyrant. The contradiction is clear, and it is intentional. It exists to keep women in a perpetual state of self-doubt, to ensure that they never lead with full authority, that they are always moderating themselves, softening their power, making themselves less threatening so that they might be accepted.

This is the trap of likeability. It is the invisible leash that keeps women in line, that ensures they are always negotiating rather than leading, always waiting to be embraced rather than commanding the space they deserve. A woman in power who worries about being liked will never lead effectively. She will be trapped in an endless cycle of justification, of seeking validation, of trying to be acceptable rather than authoritative.

And the truth is, it will never be enough. No matter how accommodating she is, no matter how carefully she moderates her tone, no matter how much she tempers her ambition, there will always be those who see her as too much. Too aggressive. Too ambitious. Too emotional. Too detached. Too something. The contradiction is not meant to be solved. It is meant to exhaust.

The most powerful women in history understood this. They did not waste time trying to be liked. They did not waste energy trying to soften themselves to make others comfortable. They ruled with the knowledge that they would never be fully accepted, and they ruled anyway.

Elizabeth I did not lead England by seeking approval. She led by making herself untouchable. She did not marry because she understood that to do so would mean ceding power, becoming a wife first and a ruler second. Instead, she created the myth of the Virgin Queen, presenting herself as a ruler who belonged only to England, beyond the reach of men who might have sought to control her. She ensured that her authority was absolute, that those who opposed her would be eliminated, that her reign would not be questioned. She did not ask for permission to govern. She governed.

Catherine the Great seized power through a coup. She knew that she would never be seen as fully legitimate, that she would always be considered an outsider, that she would always have to prove herself in ways that her male predecessors never had to. So she did not waste time trying to be loved. She focused on making herself necessary. She modernized Russia, expanded its empire, strengthened its economy. She ensured that even those who despised her had no choice but to acknowledge that she was the best ruler Russia had ever seen. She did not beg for respect. She took it. And when she faced opposition, she crushed it without hesitation.

Cleopatra did not survive in a world dominated by Rome by being soft. She ruled as a living goddess, crafting an image of herself so powerful that it became inseparable from the throne itself. She did not rely on love alone to maintain her authority. She made herself irreplaceable, ensuring that Egypt's survival was directly tied to her own. She was not just a queen. She was Egypt itself. And because of this, she was feared as much as she was admired.

But these women also understood something else: ruling through fear alone is not enough. A leader who is only feared is a leader who will eventually be overthrown. Terror without loyalty is unsustainable. A ruler who governs only through intimidation will create enemies who will one day rise against her. This is the balance that women in power must master. They must command authority without falling into the trap of cruelty for cruelty's sake. They must be respected, but they must also be necessary.

Modern women in leadership must recognize that they will always be expected to be more accommodating than men. They will always be held to a different standard, expected to be strong but not too strong, authoritative but not too authoritative, ambitious but not too ambitious. The contradiction cannot be solved, so it must be ignored. A woman who leads must not waste time adjusting herself to meet expectations that were never meant to be fair. She must decide how she will be feared, how she will be respected, how she will ensure that her authority is never questioned.

A woman who is only feared will be removed at the first opportunity. A woman who is only loved will never be taken seriously. The key is to be respected. This means setting clear expectations, enforcing consequences for disloyalty, ensuring that power is never seen as negotiable. But it also means inspiring confidence, ensuring that those who follow you do so not just out of obligation but because they believe in your vision. It means creating a legacy that extends beyond individual ambition, a sense that the leader is not just a person but a force, a necessity.

A modern woman in power must stop wasting time trying to make herself palatable. She must stop apologizing for her authority. She must stop softening herself in an attempt to avoid criticism. The criticism will come no matter what she does. She will always be seen as too much or not enough. She will always be scrutinized more harshly, judged more unfairly, held to impossible standards. The only way to win is to stop playing the game.

This does not mean embracing cruelty. It does not mean ruling through fear alone. It means ensuring that respect is never negotiable, that those who seek to undermine authority are not given the opportunity, that leadership is seen as necessary rather than optional. It means commanding authority without hesitation, without apology, without the endless justifications that women are expected to provide.

Women who lead today must study those who came before them, must understand how they maintained control, how they balanced fear and admiration, how they ensured that their authority was not just accepted but unchallengeable. They must recognize that they will never be given the same freedom to fail that men are given. They must be relentless, unshakable, immune to the demands that they be softer, more likable, more accommodating.

Because in the end, power is not about being liked. It is about being respected. And a woman who is respected is a woman who cannot be removed. She is a woman who is not negotiating for her authority but commanding it. She is a woman who is not waiting for validation but enforcing her vision. She is a woman who understands that leadership is not about fairness, that it is not about pleasing people, that it is not about being acceptable. It is about ruling.

And a woman who understands this, who embraces it without hesitation, is a woman who will win. Not because she was allowed to, not because she was invited in, but because she refused to be denied.

Part 3

Power does not belong to those who wait for approval. It does not belong to those who seek to be loved, to those who soften themselves in an effort to be accepted, to those who believe that fairness will one day prevail if they simply endure long enough. Power belongs to those who take it, who hold onto it with unrelenting force, who do not ask permission to lead but assert their authority without apology. A woman who seeks to rule must understand this. She must accept that she will never be fully embraced, that she will always be judged more harshly than a man, that she will always be expected to prove herself in ways that no male leader ever will. And she must decide to rule anyway.

Machiavelli understood the fundamental truth of power: it is better to be feared than loved. Love is conditional. It is fickle, fragile, dependent on circumstances outside of one's control. Fear, however, is reliable. It ensures obedience, demands respect, eliminates the uncertainty that comes with relying on the goodwill of others. A ruler who is feared does not have to worry about whether their subjects will remain loyal in times of crisis. Their authority is unquestioned. Their control is absolute. But for women, the balance is more complicated. A male leader who is feared is seen as strong. A female leader who is feared is seen as dangerous. A man who commands respect through force is called decisive. A woman who does the same is called ruthless. The contradiction is deliberate. It is meant to ensure that no woman can lead without being forced to justify her authority at every turn.

The trap of likability is one of the most effective tools used to keep women out of power. It convinces them that if they are warm enough, if they are patient enough, if they are accommodating enough, they will eventually be accepted. But acceptance is not the same as authority. Women who seek to be liked are rarely allowed to lead. They are seen as advisors, as supporters, as figures who exist to make others feel comfortable rather than as figures who are meant to rule. The truth is that a woman who tries to make herself palatable to everyone will never hold real power. She will be adjusting endlessly, moderating her tone, dulling her ambition, ensuring that she is always just likable enough to avoid rebellion but never authoritative enough to truly command. And this is the greatest danger of all.

A woman in power does not need to be liked. She needs to be necessary. She needs to ensure that those who follow her do so not out of affection but out of understanding that she is the only one who can lead. This does not mean ruling through terror. It does not mean being cruel for cruelty's sake. It means ensuring that respect is never negotiable, that disloyalty is met with consequences, that leadership is not seen as something that can be easily removed. A woman who is only feared will eventually face rebellion. A woman who is only loved will never be taken seriously. The key is to command both admiration and intimidation, to ensure that those who follow her understand that they have no alternative.

The women who have ruled successfully throughout history have all understood this balance. They have not wasted time seeking approval. They have not softened their leadership in an effort to make others comfortable. They have ruled with the knowledge that they would never be fully accepted, and they have ruled anyway.

Elizabeth I did not seek to be liked. She sought to be obeyed. She knew that if she attempted to lead as a woman, she would be dismissed, undermined, controlled. So she led as something more. She created the myth of the Virgin Queen, ensuring that her authority was absolute, that she was seen not as a mere mortal but as a ruler whose power was beyond question. She ensured that disloyalty was crushed without hesitation, that those who challenged her authority did not live long enough to do so twice. She was feared, but she was also admired. She ruled not because she was loved, but because she made it clear that she was the only one who could rule.

Catherine the Great did not waste time trying to make herself likable. She understood that as a foreign-born woman who had seized power through a coup, she would never be fully accepted by the Russian elite. So she did not seek their approval. She made herself indispensable. She expanded the empire, modernized its institutions, strengthened its economy. She ensured that even those who despised her had no choice but to acknowledge her effectiveness. She was feared, but she was also respected. She did not beg for legitimacy. She took it.

Cleopatra understood that power was not about fairness, that she could not afford to be seen as just another queen. She had to be more. She had to be a force, a living goddess, a ruler whose power was unquestionable. She did not rely on love alone to maintain her authority. She made herself irreplaceable, ensuring that Egypt's survival was tied directly to her own. She was not just a leader. She was Egypt itself. And because of this, she was feared as much as she was admired.

These women ruled in ways that modern women in leadership must study. They did not waste time trying to be palatable. They did not temper their ambition to make others comfortable. They did not seek approval from those who would never grant it to them. They ruled as though their authority was already unquestionable, and in doing so, they made it so.

A modern woman in power must understand that she will never be given the same freedom to fail that men are given. She will be scrutinized more harshly, judged more unfairly, held to impossible standards. And she must not let this deter her. She must recognize that every criticism, every attack, every attempt to undermine her is proof that she is doing something right. She must stop seeking validation and start enforcing her authority.

This does not mean ruling through fear alone. It means ensuring that respect is never negotiable, that those who seek to undermine her do not have the opportunity, that her leadership is seen as necessary rather than optional. It means commanding authority without hesitation, without apology, without the endless justifications that women are expected to provide.

Women who lead today must understand that they will always be expected to be more accommodating than men. They will always be expected to walk a fine line between being strong but not too strong, authoritative but not too authoritative, ambitious but not too ambitious. The contradiction cannot be solved, so it must be ignored. A woman who leads must not waste time adjusting herself to meet expectations that were never meant to be fair. She must decide how she will be feared, how she will be respected, how she will ensure that her authority is never questioned.

A woman who is only feared will be removed at the first opportunity. A woman who is only loved will never be taken seriously. The key is to be respected. This means setting clear expectations, enforcing consequences for disloyalty, ensuring that power is never seen as negotiable. But it also means inspiring confidence, ensuring that those who follow her do so not just out of obligation but because they believe in her vision. It means creating a legacy that extends beyond individual ambition, a sense that the leader is not just a person but a force, a necessity.

A modern woman in power must stop wasting time trying to make herself palatable. She must stop apologizing for her authority. She must stop softening herself in an attempt to avoid criticism. The criticism will come no matter what she does. She will always be seen as too much or not enough. She will always be scrutinized more harshly, judged more unfairly, held to impossible standards. The only way to win is to stop playing the game.

Power is not about fairness, nor is it about waiting to be accepted. A woman who understands this will not waste time trying to earn approval from those who have already decided she does not belong. She will not soften her leadership to make others comfortable, nor will she dull her ambition in hopes that the world will someday embrace her. She will rule with the knowledge that her authority will always be questioned, that her presence in spaces of power will always be seen as an anomaly, that no matter how much she accomplishes, she will always be forced to prove herself again and again. And she will rule anyway.

A woman who leads must abandon the illusion that she can avoid criticism if only she finds the right balance. She will never be enough for those who do not want her there. She will always be seen as too much or too little, too aggressive or too weak, too emotional or too detached. The contradiction is meant to keep her off balance, to keep her adjusting, to keep her spending energy trying to prove that she deserves to be in a room where men walk in without question. But a woman who understands power does not waste time playing a game she cannot win. She does not negotiate for her authority. She asserts it.

This is why a woman in power must study the women who ruled before her, the ones who refused to wait, who did not waste time proving themselves to those who would never grant them legitimacy. She must study how they controlled perception, how they wielded fear without allowing it to consume them, how they cultivated both admiration and necessity. Because power is not just about command, it is about ensuring that those who follow you understand that they have no alternative.

A woman who understands this will never rule from a place of apology. She will never lead in a way that suggests she is asking to be accepted. She will lead as though her authority is already unquestionable, and in doing so, she will make it so. Because power is not about being liked. It is about being respected. And a woman who is respected is a woman who cannot be removed. She is a woman who commands rather than negotiates, who enforces rather than explains, who leads because she has decided that she will not be denied. And that is what makes her unstoppable.

Because in the end, power is not about being liked. It is about being respected. And a woman who is respected is a woman who cannot be removed. She is a woman who is not negotiating for her authority but commanding it. She is a woman who is not waiting for validation but enforcing her vision. She is a woman who understands that leadership is not about fairness, that it is not about pleasing people, that it is not about being acceptable. It is about ruling.

And a woman who understands this, who embraces it without hesitation, is a woman who will win. Not because she was allowed to, not because she was invited in, but because she refused to be denied.

~ 9
Building Structures of Power

Beyond Individual Success

Power that does not outlast the individual who holds it is not real power. It is a temporary victory, a brief disruption in a system that will quickly recalibrate itself the moment that individual is removed. Women who seek to lead must understand this. They must recognize that their success alone is not enough, that holding a title or breaking a barrier does not mean the game has changed. If the system remains intact, if the structures that kept women out remain in place, then any power they wield will die with them.

For centuries, women have fought to gain entry into positions of power, to claim their place in leadership, to prove that they are just as capable, if not more so, than the men who have ruled by default. And time and again, they have won battles, only to see their victories reversed, their legacies erased, their positions occupied by men as soon as they are gone. This is not because they were not strong enough, not skilled enough, not determined enough. It is because they did not change the system itself. And this is the lesson that modern women in power must learn: it is not enough to succeed individually. They must build structures of power that endure beyond them, that cannot be dismantled when they leave, that ensure that they are not the exception but the beginning of something permanent.

The greatest mistake any woman in power can make is believing that her own success is enough. It is not. A woman who climbs to the top of a corporate empire, a political office, an institution, and fails to change the rules that kept her out is not a revolutionary, she is a placeholder. She may be celebrated, she may be admired, but if she has not institutionalized power, if she has not rewritten the terms of leadership, then she has done nothing to ensure that others will follow. And make no mistake: the system will close ranks the moment she is gone. It will say, "See? We let a woman in. She had her time. Now things can go back to normal."

This is why women in power must not simply focus on their own survival. They must build. They must create networks, alliances, policies, and institutions that ensure their influence extends beyond their tenure, beyond their lifetime, beyond the temporary nature of their own leadership. They must make themselves unerasable, not just as individuals, but as a force that cannot be removed.

Historically, the women who understood this did not just lead. They created systems. They laid foundations. They ensured that their power was not dependent on their individual presence but on the structures they built. Consider Eleanor Roosevelt, who did not merely serve as First Lady but used her position to institutionalize human rights as a global priority, crafting frameworks that would outlast her. Consider Ruth Bader Ginsburg, who did not merely argue for women's rights in individual cases but fundamentally reshaped legal precedents so that gender equality was not a favor granted to women but a legal standard embedded in law. Consider education pioneers like Mary McLeod Bethune, who did not just fight for Black women's education but built institutions that would continue educating generations after her death.

These women understood that personal power is fleeting. It is only as strong as the systems that sustain it. And a woman in power who does not invest in those systems is a woman who will be forgotten the moment she is gone.

This is why women who lead must think not just about their own success but about the structures that will survive them. They must build institutions that cannot be easily dismantled. They must establish networks that ensure power is not concentrated in one person but distributed so widely that it cannot be undone. They must create policies that make it impossible for future generations to be forced to start from scratch.

For too long, women in leadership have been made to believe that their role is to prove that they belong. They have been told that their success is a victory in itself, that simply holding power is enough. But this is a lie. A woman in power who does not create a system that allows other women to follow her has not truly won. She has simply occupied a space that will be taken from her the moment she is no longer there.

This is the difference between temporary victories and permanent change. The first is about individual success; the second is about institutional power. And a woman who seeks real power must stop thinking only in terms of herself. She must see her leadership as the foundation of something much larger, something that will persist even when she is no longer there to defend it.

This is why networks are essential. A woman in power cannot survive alone. She cannot assume that her personal strength, her intelligence, or her talent will be enough. The system will always try to isolate her, to ensure that she remains an exception rather than the beginning of a new order. The only way to resist this is through alliances, through the deliberate creation of networks that distribute power, that reinforce leadership, that make it impossible for the system to revert back to its original state.

History has shown that when women rule alone, they are erased. But when they build institutions, they become unstoppable. Women who have changed the world have done so not because they were exceptional individuals but because they made themselves part of something larger than themselves.

Consider the women who led the fight for voting rights. They did not merely seek to vote themselves, they ensured that the right was enshrined in law, so that no future generation would have to fight the same battle. Consider the women who built schools, who founded universities, who created legal precedents that forced society to acknowledge their rights. They did not just succeed for themselves; they created pathways so that others would not have to start from zero.

This is the work that women in power today must embrace. They must think beyond themselves. They must ensure that their leadership is not dependent on their individual presence but on the structures they create. They must build systems so strong that even when they are gone, their influence remains.

Because power that is personal is temporary. But power that is institutionalized is permanent. And a woman who understands this does not just lead, she builds. She does not just rule, she creates a new reality. She does not just fight for herself, she ensures that no one after her will have to fight the same battle again. That is the difference between individual success and real power. And a woman who understands this will never be just a leader. She will be a force that reshapes history.

But reshaping history requires deliberate action. It requires women in power to actively resist being framed as isolated figures, to refuse the temptation of becoming symbols without substance. The system loves to celebrate the lone exceptional woman while ensuring that she remains an anomaly. It loves to tell stories of the singular woman who "defied the odds" rather than acknowledging that the odds should never have been stacked against her in the first place. A woman who truly seeks to change the game must reject this narrative. She must reject the idea that she is special, that her success is proof that progress has been made, that the doors are now open for others. She must recognize that unless she forces those doors to stay open, unless she builds mechanisms to keep them from being shut again, she has done nothing but serve as a temporary exception.

This means investing in mentorship, in sponsorship, in the deliberate elevation of other women. It means understanding that power must be shared in order to be sustained. It means forming alliances, not just with those who think exactly like her, but with those who are committed to the same long-term vision of systemic change. It means ensuring that power does not begin and end with her, but flows outward, embedding itself into institutions, into policies, into the very fabric of how leadership functions.

A woman who truly understands power is never content with just holding it. She is concerned with where it will go after her, with how it will survive beyond her, with ensuring that the battles she fought never have to be fought again. Because the ultimate measure of success is not how much power she gains in her lifetime, it is how much of it she leaves behind, unshakable, untouchable, undeniable. That is the kind of power that lasts. That is the kind of power that changes the world.

Part 2

Power that does not extend beyond the individual who holds it is not real power. It is a temporary disturbance in a system designed to reset itself the moment that individual is gone. Women who seek power must recognize this truth: their personal success, no matter how groundbreaking, means nothing if it does not change the structures that kept them out in the first place. A woman who rises to power without institutionalizing that power is merely a visitor in a world that was never built for her. And visitors do not get to stay.

The system is designed to tolerate the occasional outlier, to allow a handful of women to reach high positions as long as they do not fundamentally alter the order of things. It thrives on tokenism, on the idea that if one woman succeeds, the problem must be solved, that if one woman reaches the top, the system must not be so exclusionary after all. This is how power protects itself, by creating the illusion of progress without making any real structural change. A woman who plays into this illusion, who is content with being the exception rather than dismantling the barriers that made her an exception, has not truly won. She has simply been allowed to exist within a framework that will erase her the moment she is gone.

This is why women in power must not just focus on survival. They must build. They must create networks, institutions, policies, and structures that ensure that their presence in leadership is not a fleeting anomaly but a permanent shift in how power functions. They must operate with the understanding that their work is not about their own individual rise, but about creating a system where power is not something women must fight for, it is something they inherit, control, and pass down without question.

The most powerful women in history understood this distinction. They were not content to be singular figures, to hold power in isolation, to be celebrated as rare exceptions. They built systems that endured beyond them. They ensured that their influence was institutionalized, that their leadership became a foundation rather than a footnote.

Consider Eleanor Roosevelt. She did not simply exist as a powerful First Lady, she fundamentally reshaped the role itself, expanding it into a platform for advocacy, governance, and international diplomacy. She was not content to simply advise from the sidelines. She became a force in policymaking, in human rights, in global governance. And when she left the White House, she did not fade into history. She helped shape the United Nations, building frameworks for human rights that would survive beyond her lifetime. She did not just hold power; she made sure that the institutions she influenced could not be easily undone.

Ruth Bader Ginsburg followed a similar path. She did not simply argue for gender equality on a case-by-case basis. She crafted legal strategies that embedded equality into the very structure of American law. She knew that a single victory meant nothing if it did not change precedent, if it did not force the system to evolve in a way that could not be easily reversed. She played the long game, ensuring that the victories she won were not just personal milestones but permanent shifts in the legal landscape.

Even in education, where women have historically been kept out of power, figures like Mary McLeod Bethune built institutions that ensured that access to knowledge and leadership was not dependent on the goodwill of existing systems. She did not simply demand a seat at the table, she built her own table. She founded schools, institutions, and political alliances that ensured that Black women had not just the right to education but the ability to wield power within it. She did not just fight for her own success. She fought to ensure that others would never have to struggle the way she did.

This is the lesson that modern women in power must learn. Success is meaningless if it is not institutionalized. Holding office, running a company, leading a movement, these are not victories in themselves. They are only victories if they create lasting change, if they prevent the next generation from having to fight the same battles over and over again. A woman in power who does not lay the foundation for others to follow has not won. She has only delayed the inevitable return of the status quo.

This is why networks matter. Power is not meant to be held alone. Women who lead must recognize that their greatest strength comes not from their individual accomplishments, but from the alliances they build. The system thrives on isolation, on ensuring that women in power remain disconnected from one another, that they see each other as competitors rather than collaborators. This is deliberate. A single woman in power is vulnerable. A network of women in power is unstoppable.

Historically, the greatest threats to existing power structures have come from organized movements, not individual figures. The suffragists did not win the right to vote because one woman made a compelling argument. They won because they created a network that spanned countries, generations, and classes, a force so relentless that it could not be ignored. The civil rights movement was not built on individual leaders alone, it was built on coalitions, on interconnected struggles, on an understanding that systemic change requires collective force.

Modern women in power must adopt this mindset. They must create alliances that cannot be easily broken, networks that distribute power so widely that it cannot be contained. They must recognize that their strength is not in their singular ability to succeed, but in their ability to ensure that power does not end with them.

This also means rejecting the scarcity mindset that has been imposed upon women in leadership. The system tells women that there are limited seats at the table, that they must compete with one another for a small number of opportunities, that their success is contingent on the failure of others. This is a lie. There is no inherent limit to power, only the limits that are imposed by those who seek to control it. A woman who understands this does not waste time tearing down other women. She builds with them. She expands power rather than accepting its artificial constraints.

Building structures of power also requires a deep understanding of legacy. Too often, women in leadership are told that their primary concern should be their immediate impact, that they must prove themselves in the present moment, that their value is measured by how much they personally accomplish. But the most effective leaders are those who think beyond themselves, who recognize that their greatest impact may not be felt in their own lifetime, but in the world they leave behind.

A woman who seeks real power must think in decades, not election cycles. She must think in generations, not personal achievements. She must recognize that true leadership is not about individual wins, it is about ensuring that the system itself is altered so completely that the next generation does not even have to think about fighting the same battles.

This is the difference between influence and power. Influence is fleeting. It is tied to the individual, to their reputation, to their personal ability to command attention. Power, real power, is institutional. It is embedded in governance, in law, in culture, in the very framework of how leadership operates. And a woman who understands this does not just seek to hold power, she seeks to build it.

This means thinking beyond visibility. Too often, women are encouraged to equate success with being seen, with being recognized, with being acknowledged as trailblazers. But visibility is not the same as power. A woman can be highly visible and still be disposable. A woman can be celebrated for her achievements while the structures that made her success so difficult remain unchanged. A woman who understands real power does not just seek to be visible, she seeks to be unremovable. She ensures that her influence is so deeply ingrained in the system that it cannot be undone, that it exists independently of her presence, that it survives even after she is gone.

This is why every woman who holds power must ask herself: What will remain when I am no longer here? What structures am I building that will endure beyond my tenure? What systems am I putting in place that will prevent my successors from having to start from zero? If the answer is nothing, then she has not won. She has merely occupied space temporarily, and the moment she is gone, the system will erase her as if she were never there.

The women who have truly changed history were not the ones who simply held power. They were the ones who built it, who embedded it so deeply into institutions that it could not be undone. They did not just rule for themselves. They ruled to ensure that those who came after them would never have to ask for permission to rule again. That is the work of real power. And a woman who understands this will not just lead, she will ensure that no one can take power away from those who follow.

Part 2

Power that does not extend beyond the individual who holds it is not real power. It is a temporary disturbance in a system designed to reset itself the moment that individual is gone. Women who seek power must recognize this truth: their personal success, no matter how groundbreaking, means nothing if it does not change the structures that kept them out in the first place. A woman who rises to power without institutionalizing that power is merely a visitor in a world that was never built for her. And visitors do not get to stay.

The system is designed to tolerate the occasional outlier, to allow a handful of women to reach high positions as long as they do not fundamentally alter the order of things. It thrives on tokenism, on the idea that if one woman succeeds, the problem must be solved, that if one woman reaches the top, the system must not be so exclusionary after all. This is how power protects itself, by creating the illusion of progress without making any real structural change. A woman who plays into this illusion, who is content with being the exception rather than dismantling the barriers that made her an exception, has not truly won. She has simply been allowed to exist within a framework that will erase her the moment she is gone.

This is why women in power must not just focus on survival. They must build. They must create networks, institutions, policies, and structures that ensure that their presence in leadership is not a fleeting anomaly but a permanent shift in how power functions. They must operate with the understanding that their work is not about their own individual rise, but about creating a system where power is not something women must fight for, it is something they inherit, control, and pass down without question.

The most powerful women in history understood this distinction. They were not content to be singular figures, to hold power in isolation, to be celebrated as rare exceptions. They built systems that endured beyond them. They ensured that their influence was institutionalized, that their leadership became a foundation rather than a footnote.

Consider Eleanor Roosevelt. She did not simply exist as a powerful First Lady, she fundamentally reshaped the role itself, expanding it into a platform for advocacy, governance, and international diplomacy. She was not content to simply advise from the sidelines. She became a force in policymaking, in human rights, in global governance. And when she left the White House, she did not fade into history. She helped shape the United Nations, building frameworks for human rights that would survive beyond her lifetime. She did not just hold power; she made sure that the institutions she influenced could not be easily undone.

Ruth Bader Ginsburg followed a similar path. She did not simply argue for gender equality on a case-by-case basis. She crafted legal strategies that embedded equality into the very structure of American law. She knew that a single victory meant nothing if it did not change precedent, if it did not force the system to evolve in a way that could not be easily reversed. She played the long game, ensuring that the victories she won were not just personal milestones but permanent shifts in the legal landscape.

Even in education, where women have historically been kept out of power, figures like Mary McLeod Bethune built institutions that ensured that access to knowledge and leadership was not dependent on the goodwill of existing systems. She did not simply demand a seat at the table, she built her own table. She founded schools, institutions, and political alliances that ensured that Black women had not just the right to education but the ability to wield power within it. She did not just fight for her own success. She fought to ensure that others would never have to struggle the way she did.

This is the lesson that modern women in power must learn. Success is meaningless if it is not institutionalized. Holding office, running a company, leading a movement, these are not victories in themselves. They are only victories if they create lasting change, if they prevent the next generation from having to fight the same battles over and over again. A woman in power who does not lay the foundation for others to follow has not won. She has only delayed the inevitable return of the status quo.

This is why networks matter. Power is not meant to be held alone. Women who lead must recognize that their greatest strength comes not from their individual accomplishments, but from the alliances they build. The system thrives on isolation, on ensuring that women in power remain disconnected from one another, that they see each other as competitors rather than collaborators. This is deliberate. A single woman in power is vulnerable. A network of women in power is unstoppable.

Historically, the greatest threats to existing power structures have come from organized movements, not individual figures. The suffragists did not win the right to vote because one woman made a compelling argument. They won because they created a network that spanned countries, generations, and classes, a force so relentless that it could not be ignored. The civil rights movement was not built on individual leaders alone, it was built on coalitions, on interconnected struggles, on an understanding that systemic change requires collective force.

Modern women in power must adopt this mindset. They must create alliances that cannot be easily broken, networks that distribute power so widely that it cannot be contained. They must recognize that their strength is not in their singular ability to succeed, but in their ability to ensure that power does not end with them.

This also means rejecting the scarcity mindset that has been imposed upon women in leadership. The system tells women that there are limited seats at the table, that they must compete with one another for a small number of opportunities, that their success is contingent on the failure of others. This is a lie. There is no inherent limit to power, only the limits that are imposed by those who seek to control it. A woman who understands this does not waste time tearing down other women. She builds with them. She expands power rather than accepting its artificial constraints.

Building structures of power also requires a deep understanding of legacy. Too often, women in leadership are told that their primary concern should be their immediate impact, that they must prove themselves in the present moment, that their value is measured by how much they personally accomplish. But the most effective leaders are those who think beyond themselves, who recognize that their greatest impact may not be felt in their own lifetime, but in the world they leave behind.

A woman who seeks real power must think in decades, not election cycles. She must think in generations, not personal achievements. She must recognize that true leadership is not about individual wins, it is about ensuring that the system itself is altered so completely that the next generation does not even have to think about fighting the same battles.

This is the difference between influence and power. Influence is fleeting. It is tied to the individual, to their reputation, to their personal ability to command attention. Power, real power, is institutional. It is embedded in governance, in law, in culture, in the very framework of how leadership operates. And a woman who understands this does not just seek to hold power, she seeks to build it.

This means thinking beyond visibility. Too often, women are encouraged to equate success with being seen, with being recognized, with being acknowledged as trailblazers. But visibility is not the same as power. A woman can be highly visible and still be disposable. A woman can be celebrated for her achievements while the structures that made her success so difficult remain unchanged. A woman who understands real power does not just seek to be visible, she seeks to be unremovable. She ensures that her influence is so deeply ingrained in the system that it cannot be undone, that it exists independently of her presence, that it survives even after she is gone.

The women who have truly changed history were not the ones who simply held power. They were the ones who built it, who embedded it so deeply into institutions that it could not be undone. They did not just rule for themselves. They ruled to ensure that those who came after them would never have to ask for permission to rule again.

That is the work of real power. And a woman who understands this will not just lead, she will ensure that no one can take power away from those who follow.

Power that is not institutionalized is power that will be erased. Women in leadership must recognize that their success is meaningless if it does not leave behind structures strong enough to outlast them. A woman who rises to power, no matter how high she climbs, is only a temporary disruption if she does not fundamentally change the rules of the game. She can hold office, build a company, shape policy, but if she does not create a system that ensures others like her can follow, she has merely been allowed to exist within a framework that will erase her the moment she is gone. The system will claim her as proof of progress while ensuring that the door behind her remains firmly shut.

History is filled with women who were powerful in their time but left no lasting change because they were treated as exceptions rather than as the foundation of something permanent. A single woman breaking a barrier does not mean the barrier is gone; it means it was momentarily lifted for her. If that door does not remain open, if others are not able to step through it without the same struggle, then her victory is an illusion. Real power is not about proving that a woman can lead. It is about making it impossible to return to a time when women were not leading.

This is why women in power must think beyond their own survival. They must not just take space in the world as it exists, they must change the very structure of that world so that their presence is no longer remarkable. They must understand that their power is not just about their own success but about building something that endures. If the system is not different when they leave, they have not won.

One of the greatest traps laid for women in leadership is the myth that individual success equals progress. The system loves to hold up lone women as proof that things have changed while doing nothing to alter the conditions that made their rise so difficult in the first place. It celebrates the "first" woman to achieve something, as if that single achievement means the struggle is over. But firsts are meaningless if they are not followed by seconds, thirds, and countless others who no longer have to fight the same battles.

A woman who is the first in a position of power must ensure she is not the last. She must understand that her success is a tool, not a conclusion. She cannot allow herself to be used as evidence that the system is fair when she knows it is not. Instead, she must leverage her position to dismantle the barriers that made her rise so improbable. She must institutionalize power so that it no longer requires exceptionalism for women to hold it.

Consider the women who changed not just their own circumstances but the very systems they worked within. Eleanor Roosevelt did not simply exist as a powerful First Lady, she reshaped the expectations of the role, expanding it into an instrument of governance, diplomacy, and advocacy. Ruth Bader Ginsburg did not just argue for women's rights, she built legal precedents that embedded gender equality into the foundation of the law. These women did not settle for personal influence. They changed the structure itself so that their work could not be easily undone.

A woman in power cannot afford to operate in isolation. The system thrives on keeping women disconnected, on ensuring that those who break through barriers remain alone, unable to build alliances that could threaten the established order. A woman who does not actively create networks is a woman who will be replaced the moment she is gone.

Historically, the most successful movements have not been driven by individuals alone but by networks of people committed to systemic change. The suffragists won the right to vote not because of one leader but because of a collective force that spanned generations and borders. The civil rights movement was not built on a single figure but on interconnected struggles that reinforced each other. The same is true for women in power today. They must not see themselves as isolated figures but as part of a larger force working toward something bigger than themselves.

This requires a fundamental shift in how women approach power. The scarcity mindset, where women are made to believe there are only a few seats at the table, forcing them to compete against one another, is a lie designed to keep them weak. A woman who understands power does not compete for limited space. She expands it. She ensures that power is not concentrated in one individual but spread widely enough that it cannot be easily dismantled.

A woman who builds only her career builds nothing. Her success is hers alone, and when she is gone, it vanishes with her. A woman who builds institutions, on the other hand, creates something that cannot be erased. She ensures that the structures she puts in place will continue to function even in her absence.

The most effective women in history have done exactly this. They have not merely led, they have built systems that extend beyond their own influence. Education reformers like Mary McLeod Bethune did not just fight for Black women's access to education; they created institutions that would continue that work long after their deaths. Legal pioneers like Pauli Murray did not just argue for equality in individual cases; they rewrote legal frameworks so that the fight for justice became a systemic force rather than a personal battle.

Modern women in leadership must embrace this mindset. They must ask themselves: What am I building that will outlast me? How am I ensuring that my work does not disappear when I am no longer here to defend it? What systems am I putting in place so that future generations do not have to fight the same battles?

A woman who seeks real power does not measure her success in personal milestones. She measures it in how much she changes the world for those who come after her. She knows that leadership is not about proving she can succeed within an unfair system but about making that system incapable of excluding others. She ensures that her influence is embedded in governance, law, culture, and institutions, so deeply rooted that it cannot be undone.

This requires long-term thinking. It requires women to stop focusing solely on the immediate wins and start thinking in terms of decades, in terms of generational transformation. It requires them to recognize that true leadership is not about their own victories but about making sure that the next generation does not even have to think about fighting the same battles.

A woman who plays the long game does not seek validation from the existing system. She seeks to outlast it. She builds coalitions, creates policies, reshapes institutions. She makes herself unerasable. And most importantly, she ensures that the power she builds is not dependent on her alone.

The greatest danger to a woman in power is to be remembered as an exception. The system thrives on the idea that each woman who succeeds is unique, that she is unlike the others, that her rise is proof that meritocracy exists. But meritocracy is a myth. The system is still built to exclude, and one woman breaking through does not mean that the structure has changed.

A woman who truly understands power ensures that she is not the exception but the beginning of something unstoppable. She does not settle for being the first, she creates conditions so that there is never again a question of whether a woman belongs in leadership. She does not fight only for herself, she fights to make sure no one after her has to fight the same way.

This is how real power is built. Not through individual success, but through systemic change. Not through proving that a woman can lead, but through ensuring that leadership is no longer seen as something extraordinary when it belongs to a woman.

The women who have reshaped history have not done so because they were accepted. They have done so because they refused to let power remain something that had to be fought for over and over again. They did not just exist within the system, they forced it to evolve.

A woman who understands this does not just lead. She builds. She does not just hold power. She ensures that power belongs to those who come after her. And when she is gone, she is not forgotten, because the structures she built remain, undeniable, unshakable, impossible to erase.

Special Note

The Price of Ego in Power

History is shaped not only by the women who built enduring systems of power but also by those who, despite their brilliance, failed to secure a lasting legacy. Ruth Bader Ginsburg was a formidable force who reshaped gender equality in American law, yet her refusal to step down when she could have ensured a like-minded successor has cast a shadow over her contributions. Her decision was not just personal; it was strategic miscalculation, one that has allowed a far-right judiciary to dismantle the very rights she spent her life defending. This is the danger of individual power without institutional security, progress is fragile when it is dependent on a single person rather than embedded in unshakable systems. And we are seeing the consequences unfold in real time.

It is not just that we have failed to move forward; our rights are being actively rolled back, reproductive autonomy, labor protections, voting rights, all under attack, as though the 20th century never happened. The lesson here is clear: no woman, no matter how brilliant, can assume that progress will hold if she does not take deliberate steps to cement it. Women in power cannot rely on goodwill or the illusion of permanence. They must make their legacy unerasable, not just in memory, but in law, governance, and culture, because if they don't, history will not just return to the status quo. It will erase them entirely.

Conclusion

The Future of Power is Female If We Seize It

Power will never be given freely. It has never been handed over without struggle, never been granted as a reward for patience or proof of competence. It is taken, seized, and held with unrelenting force by those who understand that waiting for fairness is a losing game. Women have been told for centuries that if they prove themselves worthy, if they wait their turn, if they are good enough, then they will be allowed into the halls of power. But history tells a different story. The women who have truly held power, who have shaped the world rather than merely surviving in it, did not wait. They did not ask. They did not seek permission. They took it. And if women today do not do the same, they will find themselves in the same cycle of exclusion and erasure that has defined their place in history for far too long.

The future of power can be female, but only if women are willing to claim it without apology. The system is not broken. It is working exactly as it was designed, to consolidate power in the hands of the few while making those on the margins believe that they can earn their way in through compliance, through perseverance, through waiting. This is the greatest trick of power: convincing those without it that if they just follow the rules long enough, they will eventually be rewarded. But the rules were not made for women to win. They were made to keep women out. The only way forward is to stop playing the game as it exists and to start creating new rules.

Women have been taught that progress comes in small, incremental steps, that they should be grateful for every slight improvement, that they should celebrate the fact that they are now allowed into spaces that were once entirely closed off to them. But representation is not the same as power. Being present in a room does not mean having control over what happens in it. If women do not dictate the terms of their leadership, they will always be at the mercy of those who do. They must stop settling for symbolic victories, for titles without authority, for inclusion that does not come with true decision-making power.

The real fight is not just about getting into power but about redefining it. It is about breaking apart the structures that make women feel as though they must play by the existing rules rather than writing their own. It is about rejecting the expectation that they must be likable in order to be effective, that they must balance strength with softness, that they must be strategic but never ruthless. Women in power are still expected to be palatable, to lead in ways that do not disrupt the status quo too much, to soften their ambition so that it does not make others uncomfortable. But power is uncomfortable. It is disruptive. It is about challenging, about forcing change, about making the world adjust to them rather than constantly adjusting themselves to fit the world.

The demand for women to be likable is one of the most effective weapons used to keep them in check. It convinces them that their power is dependent on approval, that they must not be too aggressive, too outspoken, too difficult. It convinces them that they must always be considering how others perceive them rather than focusing on their own goals. But the truth is that a woman in power will never be universally liked. No one in power ever is. A man can be ruthless and still be respected. A woman who is ruthless is demonized. But a woman who is too soft is dismissed. The contradiction is deliberate, designed to keep women constantly adjusting, constantly doubting, constantly explaining themselves. The only way to win is to stop playing.

There is no neutral ground in power. Either women seize it on their own terms, or they remain dependent on the approval of those who would rather keep them out. They must stop waiting to be invited in. They must stop asking to be accepted. They must claim authority without hesitation, without apology, without the endless justifications that they have been conditioned to provide. And they must be willing to defend it. Because the system will always try to take it back.

The question, then, is not just how women take power, but how they hold it. Because taking power is one battle. Keeping it is another. The world does not just resist women stepping into leadership, it actively works to remove them once they are there. The attacks are immediate, relentless, and often personal. The system knows that if it can make women doubt themselves, if it can make them question whether they belong, if it can make them feel the weight of constant scrutiny and criticism, then it does not need to forcibly remove them. They will remove themselves.

This is why women must go into power with a clear understanding of what it will take to stay there. They must anticipate the backlash before it arrives. They must be prepared to be called difficult, ruthless, power-hungry. They must be prepared for the double standards, for the impossible expectations, for the reality that no matter how well they lead, they will always be judged more harshly than a man in the same position. And they must decide to lead anyway.

But this is not just about surviving in power. It is about transforming it. Women who seek power must go beyond merely existing within the structures that have historically excluded them. They must change those structures. They must redefine what leadership looks like, what it means to wield authority, what it means to build something that cannot be dismantled the moment they are gone. This is not about proving that women can succeed in the existing system. It is about making sure that future generations do not have to prove anything at all.

And this leads to the bigger question: what does it mean to rule justly in an unjust world? Because power itself is not neutral. It is shaped by the forces that have historically controlled it. The very definition of leadership has been written by those who have held it the longest. Women stepping into power must decide whether they will play by those same rules or whether they will redefine the game entirely. They must ask themselves whether justice can ever come from a system that was not built for it.

The truth is that power must be taken, but once it is taken, it must be used differently. Women cannot simply replicate the leadership models that have existed for centuries and expect a different outcome. They cannot wield power in ways that uphold the same inequalities they fought to overcome. They must be intentional about how they govern, how they lead, how they build something that is not just about them but about those who come after them.

This is where the challenge lies. Women must be ruthless in their pursuit of power, but they must also be architects of something greater. They must not just fight to exist in leadership. They must transform leadership itself. They must ensure that power is not just held by women, but that it is shaped by them, that it reflects the values and structures that truly create justice rather than just enforcing a different version of the same hierarchy.

This is the future of power, if women are willing to seize it. If they are willing to lead without apology, without waiting for permission, without softening their ambition to fit the expectations of a world that never wanted them to lead in the first place. The greatest threat to the existing order is not just women in power, but women who understand that they do not have to play by the old rules. Women who create new ones. Women who refuse to be contained.

Power is not something to be asked for. It is not something to be waited on. It is not something to be given. It must be claimed. And once it is claimed, it must never be given back. Because the world does not need more women proving that they deserve to lead. It needs women who lead as though there was never any question that they would.

Special Note

For the Builders, the Creators, and the Future

Burn the old world down. Not with fire, not with riots, not with desperate pleas for justice from the very systems that exist to deny it. Burn it down by making it irrelevant. By making it obsolete. By making those who hoard wealth and power panic at the realization that they have lost control, not because we stormed their gates, but because we built something beyond their reach.

Women have been told to climb ladders that lead to nowhere, to break glass ceilings in buildings that should have been demolished long ago. They are told that power is in politics and business, in governance and finance, in playing the game better than men do. That is a lie. The real power, the kind that cannot be taken away, cannot be voted out, cannot be bought or sold, has already begun shifting. It is not in the rotting halls of government. It is not in the hands of the billionaires strip-mining the planet. It is not in the Fortune 500, where the most ambitious women are handed empty titles while the real decisions are made behind closed doors.

The future of power belongs to those who refuse to participate in this dead empire. The writers, the artists, the coders, the creators, the builders of new systems, those who shape culture, control narratives, and develop technologies that eliminate the need for permission. Decentralized economies, unregulated communication networks, underground movements that answer to no one. Power is not taken from the old world, it is extracted from it, removed entirely, and reconstructed elsewhere.

Understand this: anything centralized can and will be controlled. Apps, social media platforms, bank accounts, even the illusion of free speech, all of it can be erased with a single keystroke. Political careers end in disgrace, corporate success is chained to investors and shareholders, and everything owned by a single entity can be bought, censored, or dismantled.

But what cannot be erased? Bitcoin. Blockchain. AI beyond corporate control. Decentralized systems where power is not granted, it is coded into existence and untouchable by those who think they still rule. Music, literature, art, and philosophy that do not seek approval from the dying order but instead create the world that will replace it.

The revolution will not be televised. It will be encrypted. It will be coded. It will be built.

Do not waste your life climbing their ladders. Do not beg for a seat at their table. Do not waste another moment asking corrupt systems to make space for you. Destroy their relevance. Build your own networks. Seize control of culture, of information, of the means to survive outside their grip. Make them obsolete.

They are already afraid. The wealth hoarders, the political puppets, the CEOs playing god while their companies cannibalize the planet. They know the shift has begun. Their power is dependent on your participation. The second you stop believing in their system, it collapses.

So stop. Stop feeding the machine that grinds you down. Stop waiting for a dying empire to change. Stop hoping they will let you in when you were meant to build something they can never enter.

This is your moment. Take it.

About EATMS Productions

What's happening to women now is not random. It's structural.

Policy, culture, technology, and power are moving in the same direction.

EATMS maps them clearly and shows how to respond.

This title is part of an ongoing body of work. All EATMS Productions titles, across all series, authors, and formats, are components of a single connected project.

Start here: EATMS System Primer — Free Bundle
https://eatms.gumroad.com/l/dyvzbw

For full catalog or inquiries: eatms.me

Free survival booklet + EATMS updates: email "EATMS" to eatms@pm.me

Please feel free to burn part or all of this book, safely, as an effigy.

www.ingramcontent.com/pod-product-compliance
Lightning Source LLC
LaVergne TN
LVHW051004080826
845145LV00009B/2449

* 9 7 8 1 9 6 6 0 1 4 1 2 6 *